Law, Ethics and the Biopolitical

Law, Ethics and the Biopolitical explores the idea that legal authority is no longer related to national sovereignty, but to the 'moral' attempt to nurture life. The book argues that while the relationship between law and ethics has long been a central concern in legal studies, it is now the relationship between law and life that is becoming crucial. The waning legitimacy of conventional conceptions of sovereignty is signalled in the renewal of a version of natural law, evident in discourses of human rights, that de-emphasises the role of a divine lawgiver in favour of an Aristotelian conception of the natural purpose of life and the 'common good'. Synthesising elements of legal scholarship on sovereignty, theories of biopolitics and biopower, as well as recent developments in the domains of ethics, Amy Swiffen examines the invocation of 'life' as a foundation for legal authority. The book documents the connection between law, life and contemporary forms of biopolitical power by critically analysing the fundamental principles of the bioethical paradigm. Unique in its critical and cross-disciplinary approach, *Law, Ethics and the Biopolitical* will be of interest to students and teachers in the areas of law and society, law and literature, critical legal studies, social theory, bioethics, psychoanalysis and biopolitics.

Amy Swiffen is an Assistant Professor of Law and Society in the Department of Sociology and Anthropology at Concordia University, Montreal, Canada. Her research is focused on new legal contexts and has been published in a wide variety of academic and professional journals.

Law, Ethics and the Biopolitical

Amy Swiffen

ROUTLEDGE
Routledge
Taylor & Francis Group
a GlassHouse book

First published 2011
by Routledge
2 Park Square, Milton Park, Abingdon, Oxon, OX14 4RN

Simultaneously published in the USA and Canada
by Routledge
270 Madison Avenue, New York, NY 10016

A GlassHouse book

Routledge is an imprint of the Taylor & Francis Group, an informa business

Typeset in Times New Roman by Taylor & Francis Books

British Library Cataloguing in Publication Data
A catalogue record for this book is available from the British Library

Library of Congress Cataloguing in Publication Data
Swiffen, Amy.
Law, ethics and the biopolitical / Amy Swiffen.
p. cm.
"A GlassHouse book"
1. Law and ethics. 2. Natural law. 3. Sociological jurisprudence.
4. Common good. I. Title.
K247.6.S93 2011
340'.112–dc22
2010025755

ISBN13: 978-0-415-57844-8 (hbk)
ISBN13: 978-0-203-83475-6 (ebk)

Contents

Introduction

This book is a consideration of the relationship between law and morality in new forms of ethical and political thought emerging in light of challenges to the legal and moral authority of national sovereignty. Notable in this respect is the rise in the last half of the twentieth century of bioethics, an ethical paradigm derived from medical ethics and born in the twentieth century. Bioethics appeals to the idea of a 'common morality' based on the universal desire for survival and value of life. An ethics based on life is seen as a natural morality (Cascais 2003) and a form of ethical thinking that can encompass all moral problems (Gert *et al.* 2006: 4). As a result, the relationship between bioethics and political concepts has yet to be considered. Another notable development that will be explored is a renewed interest in natural law theory, which is an approach to law that defines its authority in terms of universal principles found in nature, not political institutions or social conventions. Both of these are significant developments in cultural contexts where, until recently, universal principles have been regarded with incredulity, and faith in the ability of reason to lead universal maxims and the discovery of valued social goods is lacking (Pavlich 2007).

The present work draws on alternative theoretical paradigms, which, in different ways, denaturalise the idea of life. The ethics of psychoanalysis are drawn upon to elaborate an analytical framework for scrutinising the position of enunciation behind the invocation of the value of life. Theories of biopolitics are then taken up to elaborate an understanding of life as a politically determined concept beyond the good. Three literatures are drawn on in order to undertake this analysis. The first is the ethics of psychoanalysis developed by Jacques Lacan (1992), which will be developed in the first two chapters to provide an alternative ethical framework for looking at morality, specifically notions of goodness. The second is political theory, where concepts of biopolitics and biopower are playing an increasing role in debates on law and morality, largely because they are focused on power at the level of the life and death of legal subjects. The third is selected strands of legal philosophy: natural law theory and legal positivism. It is worth outlining briefly each of these fields in the context of the present analysis.

The ethics of psychoanalysis

In the opening lecture of the seminar on the ethics of psychoanalysis, Lacan invokes a distinction between morality and ethics, and explains that his choice of the latter term is precise: 'I chose a word which to my mind was not accident … If I say "ethics", you will soon see why. It is not because I take pleasure in using a term less common' (1992: 2). The concept of morality is a term for a focus on the relationship between the subject and its action in relation to 'a direction, a trajectory, in a word, a good that appeals to … an ideal of conduct' (Lacan 1992: 3). Ethics, on the other hand, is not directly concerned with the good but with the relationship between the action of a subject and their desire (1992: 219). Thus, in choosing to discuss the 'ethics' of psychoanalysis in contrast with 'morality', Lacan is emphasising the idea that while morality is premised on an ideal of the good, in order to consider ethics from the point of view of desire 'a radical repudiation of a certain ideal of the good is necessary' (1992: 230). Lacan invokes Aristotelian ethics, using it as a contrast point, and suggests that his own position can be seen in sharp relief in comparison: 'Perhaps the question will only be seen in sharp relief, when one compares the position … with that which is, for example, articulated in the work of Aristotle in connection with ethics' (1992: 5).

Aristotle's major contribution to the domain of ethics is found in a text called *The Nicomachean Ethics*. He opens the discussion with the premise that every action is undertaken with some end or purpose, no matter how basic: 'every skill and every inquiry, and similarly, every action and choice of action is thought to have some good as its goal' (1999: 1.1). He therefore defines the idea of the good as the goal of all human action. What is the idea of the good for human beings? In the most general sense, the good is happiness. Aristotle felt that no matter what we are doing we are ultimately always trying to achieve happiness, even if we do not always seek it out in the most prudent way. Furthermore, he reasoned that the experience of pleasure is what makes human beings happy; thus, if happiness is a universal good, the guide to happiness is pleasure. Thus, happiness is the highest good in Aristotle's morality and the path to the good is blazed by pleasure.

However, true happiness is not attainable by just any kind of pleasure; it is blazed by rational pleasure. The use of reason in satisfying the desire for pleasure is essential to achieving happiness according to Aristotle. Living a happy life involves harmonising our natural inclination towards pleasure and our natural ability to reason. The method for doing this is often referred to as the golden mean. Aristotle defined the golden mean as a path between the extremes of excess and lack in any given matter. Thus, it would be accurate to say that this ancient morality is based on living according to our nature as rational and pleasure-seeking animals. The idea of happiness is the ideal reference point for the golden mean and living in accordance with the satisfaction of desires in a rational and moderated way. Living in such a fashion engenders a state of happiness or the 'good life'.

What I want to highlight in this framework is how the moral connection between pleasure and rationality is derived from a prior division between different forms of pleasure and delimitation of unnatural pleasure (Lacan 1992: 3). The division in question is mentioned briefly amid a much longer discussion of pleasure and how vice and virtue are different attitudes towards pleasure. Aristotle begins with distinguishing three forms of badness. The one that is relevant in this context is called brutishness and is different from the others. The other two are degrees of deviation from the golden mean, called incontinence and vice respectively. The issue with brutishness is not how much it deviates from the golden mean, but the nature of the pleasure involved in the first place.

Simply put, 'brutish' pleasure involves enjoying things that are not naturally pleasurable. Things that are naturally pleasurable are those that stimulate the activity of a given nature, which is why humans enjoy rational activity; it is stimulating to our capacity for reason (Aristotle 1999: 1148b15–1149a2). Those pleasures that are not naturally pleasurable are ones that cannot lead to happiness and to which the golden mean cannot apply. Aristotle's examples include cannibalism, women who eat foetuses, nail biting, and hair pulling. He explains that brutishness comes about in three ways. One, an individual might have an innate deformity in their nature; the modern idea of the psychopath is probably akin to what Aristotle was imagining. Second, through acquired habit, and by this he probably was referring to abusive treatment in childhood (Sihvola 2002). Third, a brutish nature arises from 'sickness' such as a fever (Aristotle 1999: 1149a2–24). Regardless of the origin, however, the point is that a brutish nature comes about naturally, not from bad moral choices. Thus, for Aristotle not only is brutishness not morally blameworthy, it is not even part of the moral domain. The idea of happiness can thus be seen as a limit and threshold beyond which a 'whole register of nature … is literally situated by Aristotle outside the moral field' (Lacan 1992: 5). From the point of view of the ethics of psychoanalysis, this is 'surprising, primitive, paradoxical, in truth, incomprehensible' (ibid.).

Thus, one of the major themes of the ethics of psychoanalysis is the way that the ideal of the good (such as happiness) presents with a seeming naturalness but can also be conceived in a non-moral sense. If considered from the perspective of desire, the good can be seen as a threshold and limit. Thus, the ethics of psychoanalysis is not centred on the good or concerned with determining the value of action in relation to an ideal, but in the relationship between action and the desire that inhabits it (Lacan 1992: 8). In other words, it is based on a non-moral perspective:

> the ethics of psychoanalysis – for there is one – involves effacement, setting aside, withdrawal, indeed, the absence of a dimension that one

> only has to mention in order to realise how much separates us from all ethical thought that preceded us. I mean the dimension of habits, of good and bad habits.
>
> (Lacan 1992: 10)

At one point, Lacan puts the issue to the audience in terms of the clinic and the cure as an ideal of the good. In contrast to the desire to cure, which Lacan calls a benevolent fraud, the ethics of psychoanalysis is based on a 'non-desire to cure' (1992: 219). In the absence of the normative dimension to adjudicate good and bad, the question of ethics becomes: 'Which good are you pursuing precisely as far as your passion is concerned?' (Lacan 1992: 218). This perspective is completely new in ethical thought and even a 'reversal' of the Aristotelian morality (1992: 13). At minimum, it opens up a dimension of meaning and action to ethical consideration that is occluded if we conceive of ethics in terms of the good. The good, once seen to function as a limit, also becomes a threshold of power.

The morality of law

The issue of morality and legal authority has been discussed at various points in the history of legal theory. The debate that crystallises the issues at stake in this discussion began in the mid-twentieth century and the protagonists were Harvard professors H.L.A. Hart (1961) and Lon Fuller (1969). At the time, it was the catastrophe of Nazism and the acquiescence of German jurists that figured as the background for their discussions. The heart of their disagreement was whether a law can really be considered *law* if it is supporting a morally reprehensible political form. On the one hand, Hart's 'positive law' perspective was that morality is a subjective judgement; definitions of justice and goodness are normative in the last instance, so to what degree a form of law approximates morality can only be decided from a point of view internal to a legal order, not from an objective outside. Thus, from this perspective law is a matter of purposive activity while morality is concerned with a judgement about the value of purposes themselves. According to the positive concept of law, identifying when law is present is distinct from saying that it ought to be obeyed, which is another way of saying that it has moral force. Thus, Hart formulated the 'separability thesis', which states that law and morality should be separated in legal analysis (Schauer 1996). One should clearly distinguish the conditions of legality from the conditions of morality and leave the latter to ethicists. According to the separability thesis, there are no necessary moral limits on legal validity.

Indeed, one of the hallmarks of the positivist approach is to believe that 'legal norms can have any kind of content and be valid' (Kelsen 1945: 113). The idea of 'validity' here relates to the status of laws inside a legal system, not to their status relative to general or universal moral standards. For example, from the positivist perspective, legal violence is valid if it is exercised within the bounds of legality as defined in a given order (Benjamin 1968). This suspends the question of whether a particular legal system is morally good or bad. Suspending morality means that legal analysis may potentially go beyond the limit of the good and view law scientifically as a form of rational activity. Thus, while morality is subjective, the existence of law is objective. Indeed, legal positivism has been 'mainly concerned to produce a rich and accurate *description* of law and legal systems as they function in human societies' (George 1996: 339).

Fuller's counter to this was the 'overlap thesis', which states that law and morality are not dissociable; moreover, some overlap with morality is necessary for a rule to qualify as law at all. The naturalist view is that legal authority comes from human beings' own ability to recognise objective moral principles, which ostensibly exist and relate to justice. The idea is that rational reflection reveals an objective moral order that is not metaphysical but basic to nature (Finnis 1980). This was a 'natural law' approach insofar as it defined law as having a moral purpose of its own that does not depend on social norms or political institutions. Thus, from the naturalist point of view, it is fair to say that humans do not create law so much as discover it. And, the purpose of law is to instantiate principles conducive to the common good of a community (Murphy 2006: 3). Law based on such principles is binding in human societies because of our common human nature. Law that is counter to this morality is not valid. Legal violence is justified when it is a means to the ends of this morality (Benjamin 1968). Hence, the natural law quip that an unjust law is no law at all.

What is at stake in the different conceptions of the connection between law and morality can be effectively illustrated by Hart's review of Fuller's book called *The Morality of Law* (1965). In his review, Hart essentially aims to show that Fuller's natural law approach obscures the operative difference between legality and morality, and he uses the analogy of poisoning to make the point. He writes that poisoning 'is no doubt a purposive activity and reflections on its purpose may show that it has its internal principles of better and worse poisoning' (Hart 1965: 1285–86). In other words, rational reflection upon the art of poisoning allows for the elaboration of rules that members of a community of poisoners might agree on to distinguish between good and bad poisoning (for example: do not administer poisons that cause a slow death, do not administer poisons that cause vomiting, etc.). The analogy is meant to show that recognising *legal* validity in a society of poisoners is distinct from having the belief that poisoning itself is a morally correct organising principle for a legal order. The point is that defining law as a moral concept hampers the capacity to make a distinction between the *desire* for efficiency internal to a purpose (such as good poisoning) and a value judgement about the purpose itself (Hart 1965: 1285–86). In this sense, poisoning is meant to be analogous to any particular political ethos elevated to a universal purpose of law. Hart even suggests that the particular morality of law that Fuller describes actually corresponds to a liberal democratic ethos, and values of autonomy and personhood connected to it (Allan 2001).[1]

The heart of the issue in the debate is the source, scope and legitimacy of the authority of law. The naturalist view holds that 'positing a law and recognising something as law are not distinct activities' (Finnis 1996: 205). Thus, the complete meaning of the overlap thesis is also a rationale for the coercive aspect of law. This moral intuition is a hallmark of the naturalist approach. The idea of an overlap between law and morality justifies the force of law in, for example, the criminal law power of the state. Natural law theories argue that the morality they draw on is not particular but universal, and thus, that the force of law essentially comes from an extra-legal referent. The force of law derives from the fact that law is necessary for human survival. Given the nature of human beings and their circumstances, the existence of legal authority is what enables people to live together in communities (Braybrooke 2001: 125). This seeming fact of nature is the source of legal validity from the naturalist perspective. In this way, the idea of a 'natural' relationship between law and morality is also a meta-ethical explanation and justification for the 'obligatory force' of law (in the fullest sense of 'obligation') (Finnis 1980: 23–24).

Indeed, one difficulty for the positivist approach in recent years lies in the fact that in place of an extra-legal support, the concept of sovereignty has historically functioned as a de facto and apparently analytically neutral explanation for legal authority (George 1996). The concept of sovereignty is most fundamentally defined in terms of two properties. First, sovereignty refers to the highest authority or power in a legal order. Something is sovereign when there is no external limit on what it can do and no higher power that can check its actions. This is one essential attribute of sovereignty; it is said to be 'unlimited'. The power of the king in a monarchy would be one example of unlimited sovereignty, as was the status of the nation state in international law prior to the twentieth century. The second defining property of sovereignty is that it is the only force that can make laws within a given jurisdiction. Something is sovereign when there are no competing or equivalent powers inside its territory. For example, a state is sovereign within the borders of a nation and sovereignty is the force that maintains the rule of law at the national level. So, this is the second defining feature of sovereignty; it is also an internally indivisible power, which is one reason why pulls for sovereignty by sub-state groups are so resisted by national governments.

Eighteenth-century political theorists John Austin and Jeremy Bentham are generally regarded as the first to articulate sovereignty along these lines (Veitch *et al.* 2007: 12). Their so-called command theories of law rejected the theologically informed naturalist accounts of legal authority circulating during the eighteenth century, and turned instead to a conception of sovereignty to explain its existence (Austin 1954; Bentham 1977). Both saw it as an exclusive property of the state whose existence in perpetuity was a plain political fact or, as sociologists would tend to say, a social fact, which nicely explained both the source and the scope of law. The essence of the social fact thesis is the premise that legal authority is self-evident in the person or assembly of persons that people in a given territory habitually obey, for whatever reason (MacCormick 1996). The concept of sovereignty is therefore also a point beyond which inquiry into legal authority is not possible. In legal theory, sovereignty thus functioned as a quasi-universal meta-concept limit beyond which legal inquiry serves no practical purpose (Capps 2006: 20; Waldorn 1991).

Hart was addressing this issue when he elaborated the difference between perpetuating sovereignty by threat of violence and perpetuating sovereignty by the acceptance of a rule of recognition. The latter feature, he argued, was a part of a veritable and legitimate form of law, while the former was a characteristic of a deficient legal system. The difference involves the existence of what he termed 'secondary' rules. Primary rules govern behaviour and conduct of legal subjects while secondary rules set the terms for the creation and adjudication of the primary laws. Secondary rules are meta-laws that have as their content the primary laws themselves. As Hart explains, they are:

> said to be on a different level from the primary rules, for they are all about such rules; in the sense that while primary rules are concerned with the actions that individuals must or must not do, these secondary rules are all concerned with the primary rules themselves. They specify the way in which the primary rules may be conclusively ascertained, introduced, eliminated, varied, and the fact of their violation conclusively determined.
>
> (Hart 1994: 92)

Unlike the command theories, Hart's concept of law did not base legal authority on the fact of sovereignty alone, but on the existence of a social consensus on secondary rules for its recognition. If the 'recognition' is lacking a law cannot be said to be valid (Hart 1994: 50–61).[2] The force of law depends on the existence of such a 'general social practice' for recognising the right to make law (Hart 1994: 54–55). In this way, Hart retained the social fact thesis by understanding legal authority as normative in the last instance.

This response was an important development in legal positivism because by understanding legal authority as based on rules commonly accepted by those that follow them, Hart's work was seen to have quelled the naturalist 'myth' that legal authority comes from something extra-legal, such as nature, reason or violence (MacCormick 1996: 178). The positivist concept of law instead defines law in terms of 'the union of primary and secondary rules' (Hart 1994: 107). If it is acknowledged that sovereignty is an internally generated property constituted by the acceptance of rules, one of the implications is that sovereignty cannot intelligibly be regarded as the source of all the laws that make up a legal system (Waldorn 2006: 1701). Thus, it is possible to entertain the idea that a legal system can be constituted in ways that do not involve a state. Hart's account opens up the possibility that a legal system need not have state-based sovereignty at its base.

The reason I focus on these philosophical approaches to law is because of the way that the politics of law is bracketed by the assumption of social consensus.[3] For example, neither the positivist nor naturalist perspective justifies the authority of law in principle; in the last instance both defer to criteria for cases of its use. That is, neither approach can answer the question of whether legal violence 'as a principle, can be a moral means even to just ends' (Benjamin 1968: 278). From the naturalist perspective, the question is answered by the justness of the morality to which law naturally refers and uses as a criterion for making distinctions between 'good' and 'bad' cases of the use of violence. In this sense, the conventionality thesis is also operative in the naturalist view at a formal non-normative level (i.e. as an ontological fact). From the positivist perspective, the normativity of law provides the answer in specific cases. A social fact (of consensus) determines the standard of legality in a given legal order.

However, in the context of law, the maintenance of consensus depends on disciplining behaviour destructive to the common good. Maintaining standards of proper behaviour is not separable from the use of violence against those who do not share the common vision (Finnemore 2003). A legal order that regulates behaviours destructive to the common good is at the same time using coercion against those who do not share that morality. In the context of international law, for example, many argue that there are international legal norms but there is no consensus as to what they are (Goldsmith and Posner 2005: 201). This is why, for example, doctrinal analyses alone that show the legality of coercion by international law in the name of human rights vis-à-vis state sovereignty are not adequate to generate the force of law.

According to Hart, the authority of secondary rules comes from the existence of a consensus among officials to regard them as valid and govern their behaviour in accordance with them. This is known as the conventionality thesis. It is a version of what is known in sociological theory as the consensus view of law. The consensus thesis defines society by common interests and shared norms, which the law then confirms. While this thesis has been displaced in many ways in sociological theory, in legal studies social consensus is still often assumed. It shifts the problematic of the rationale for legal authority on to normativity. However, this does not avoid the problem. The issue remains of the force behind secondary rules: why do rules come to be commonly accepted at all? In positivism, the role of consensus is contingent, yet still necessary to the concept of law. The authority of law is justified by its status with regards to a consensus on the meaning of legality within a given order.

In this regard, it is worth recalling that even die-hard positivist Hart granted that there is a minimum content of natural law in all forms of legal order: what he called the natural law of survival (1958: 593). Several new natural law thinkers, such as Finnis (1980), Fuller (1969) and Murphy (2006) have turned to conceptions of morality in nature to develop a natural law theory that does not depend on a notion of the divine or of sovereignty. Hart's widely respected emphasis on life as the only admissible natural purpose of law seems to provide additional testimony to the intuitive acceptability of the value of life (as, for example, in discourses of human rights) as an objective morality basic to nature that exists over and above particular political institutions and cultural norms (Kainz 2004). As he writes:

> What makes sense of this mode of thought and expression is something entirely obvious: it is the tacit assumption that the *proper end* of human activity is survival, and this rests on the simple contingent fact that most men most of the time wish to continue its existence.
>
> (Hart 1994: 187)

Moreover, with the formulation of a natural law of survival underlying the tendency to develop legal systems in the most generic sense, Hart offered the possibility that a legal system can survive arguably even the loss of national sovereignty and remain legitimate in some sense. The sense of legitimacy derives from an appeal to conventionality like the original command aspect of positivist theories of law that Hart rejected. For it is seemingly obvious that human beings normally desire life and continued survival, and life is therefore the basis for, at least, a nearly universal consensus. Thus, a common morality based in nature subtends the very concept of law and legal authority in all human societies and can be a natural motivation for the obligation to obey the law in any political setting.

Thus, while sovereignty has long been tied to the legal positivist side of things, and seen as a meta-ethically neutral concept which marked the source, scope and authority of law in a legal order (with or without some moral system), the attempts to define legal authority first as a social fact based in a population's 'habits' of obedience and later in the acceptance of secondary rules, nonetheless relied on some conceptions of a natural tendency to form political orders. Such a tendency 'is latent in our identification of certain human needs which it is good to satisfy' (Hart 1994: 186). Thus, even Hart's attempts to define sovereignty nonetheless relied on some conceptions of a 'natural' tendency to form political orders. Therefore, it makes sense to expose and further articulate that order itself rather than relying on a sovereign solution that is proving not to be permanent.

There are two issues that will be addressed from these debates in this study. The first is the relationship between nature and reason. And second is life as the form of the good, in particular how the consensus it implies is not simply found in nature, but is rather a politically determined concept. These are issues that have not been considered in legal philosophy but they are points where new theories of biopolitics begin.

The biopolitical

Michel Foucault opens his study of the operation of power in modern society in *Discipline and Punish: The Birth of the Prison* (1977) with an account of the torture and execution of Robert-François Damiens. Damiens was condemned by the Parliament of Paris in 1757 for a plot to kill King Louis XV of France (3–8).[4] The case exemplifies how sovereignty was traditionally constituted through public displays of an excessively negative power that used extreme violence on the body of an individual to the point of death. Foucault mentions the incident in order to contrast the operation of power evident in Damiens' horrifying execution with the finely detailed and organised routines of life in nineteenth-century prisons. The two examples juxtaposed are meant to throw into relief another dimension of the operation of power in society, a new and positive aspect that is productive and creative. Foucault was highlighting mechanisms of power that are based less on public displays of violence and more on control and management of populations (Foucault 1978: 89).

In Foucault's analysis, classic sovereignty operated like 'a means of deduction, a subtraction mechanism' that imposes levies on subjects and manifests in the 'right to appropriate a portion of wealth, a tax of products, goods and services, labour and blood' (Foucault 1978: 136). It is most evident in the legislative, prohibitive and censoring dimension of law, which will ultimately kill if it is threatened, but normally allows living. In this sense, sovereignty is the right to take life or let live, and its ultimate symbol is the sword (Ojakangas 2005: 6). This form of power was based on the enactment of violence that ceased hold of life in order to suppress it (Foucault 1978: 136). Describing how such public executions were already in decline when Damiens was executed, Foucault's subsequent study describes how the operation of power of law changed with the transition from monarchy to constitutional rule in Europe from the seventeenth century onwards. Under classic sovereignty, law was considered an extension of the body of the sovereign, and so the punishment for attacking the integrity of the legal order is no different from attacking the body of the sovereign. Thus, under this form of law, the sovereignty of the law enacts revenge upon the offender's body for having been 'injured' by a crime. Classical sovereign power manifested in localised and negating displays of violence (Foucault 1978: 94–95).

The reason for discussing Damiens' execution is to exemplify the public executions characteristic of classic sovereignty that quickly became – in the context of the needs of modern state and industry – an ineffective use of the body by the law; it was not only too haphazard, but also contradicted the techniques of discipline that were emerging to create 'docile bodies'. The open displays of violence associated with classical sovereignty that negated life began to elicit outrage and sympathy from the public for the body that was the target, signalling the retreating efficacy of such displays. Humanitarian challenges to the classic operation of sovereign power came hand in hand with the emergence of new political problems related to the health, housing, habitation, and living conditions of populations (Dean 2004: 19). This is part of a process where new techniques of power emerge that centre on the administration of life, rather than the power over death.

In *The History of Sexuality* (1978), Foucault describes more specifically how in the eighteenth century the localised manifestations of sovereign violence declined as legal authority became based on the sovereignty of nation states, the power of which depended on the life and health of their populations more than the fear of punishment. As the health of populations became a primary concern for the state, increasingly the unilateral and negating aspect of sovereignty makes room for a form of 'biopower' that also becomes a political force. Biopower is the name for power that is fused with life and concerned with fostering life. Since the eighteenth century the power of life has become an increasingly important component of politics and law (Foucault 1977: 194). On the one hand, it involves the regulation of population-level phenomena such as birth, death, sickness, disease, health, sexual relations, etc., and on the other, disciplining individual bodies in institutionalised settings, through techniques that were first cultivated in prisons, the military, hospitals, factories and schools (1978). These techniques were gradually applied to social regulation and management more broadly in order to foster the usefulness and docility of individual bodies and populations (Foucault 1978: 138–39; 2007: 62). Thus, whereas classic sovereign power subtracted life, biopower is productive and creates forms of life.[5] In this sense, it is internal to a social body, relational, circulating, capillary, unpredictable and a matter of knowledge, not an external intervention that destroys life and brings death.

Unlike the localised displays of violence, these techniques circulate throughout every life in the social particularly in institutional settings such as factories, hospitals and schools. Biopower is a:

> new method of power whose operation is not ensured by right but by technique, not by law but by normalization, not by punishment but by control, methods that are employed on all levels and in forms that go beyond the state and its apparatus.
>
> (Foucault 1978: 89)

These techniques of power are not repressive, nor do they have a central point from which they emanate; it is exclusively focused on generating forces, making them grow and ordering them (1978: 136). Therefore, in the context of biopower, law becomes associated with the management of life and keeping death, if not at bay, at least under control; in this sense, death figures as a limit of a form of power that seeks to control life. Legal authority in post-monarchic, state-centred contexts comes to have a stake in the life of subjects and is concerned less with establishing an absolute limit than with normalising:

> In the course of the eighteenth century the establishment of an explicit, coded and formally egalitarian juridical framework, was made possible by the organization of a parliamentary, representative regime. But the development and generalization of disciplinary mechanisms constituted the other, dark side of these processes. The general juridical form that guaranteed a system of rights that were egalitarian in principle was supported by these tiny, everyday, physical mechanisms, by all those systems of micro-power that are essentially non-egalitarian and asymmetrical that we call the disciplines.
>
> (1977: 222)

While initially biopower and sovereignty seem opposed, much in the same way that life and death seem opposed, they are in fact not mutually exclusive. The censoring, negating, life-extinguishing aspect visible in the displays of classic sovereignty is now one element among others in a power bent on making forces grow, not destroying them (Foucault 1978: 136). This analysis identifies a profound shift in the operation of power in the form of law in society.

At moments it may seem that Foucault overstates the shift from one to the other, especially in the introduction to *The History of Sexuality*, but fundamentally the process he is referring to is one of overlap and interconnection. Sovereignty has gradually been 'penetrated' by new mechanisms of power that took charge of human existence, specifically the existence of the living body (Foucault 1978: 88). Biopolitics emerges when the ultimate reference point for political authority is no longer the threat of death but the very biological existence of the population (Foucault 1978: 137).[6] A stark example is the scale of modern warfare:

> As the technology of wars have caused them to tend increasingly toward all-out destruction, the decision that initiates them and the one that terminates them are in fact increasingly informed by the naked question of survival … If genocide is indeed the dream of modern powers, this is not because of a recent return of the ancient right to kill; it is because power is situated and exercised at the level of life.
>
> (Foucault 1978: 137–38)

In other words, the death that was based on the right of sovereignty is now manifested as the inverse of the right of the social body to ensure, maintain or develop its life (Foucault 1978: 136):

> The state has essentially to take care of men as a population. It wields its power over living beings as living beings, and its politics, therefore has to be a bio-politics. Since the population is nothing more than what the state takes care of for its own sake, of course, the state is entitled to slaughter it. So the reverse of bio-politics is thanato-politics.
>
> (Foucault 1988: 160)

Thus, the era of biopower is not a transition away from the power of death. Rather, biopower describes the ability to guarantee an individual's continued existence on the basis of the power to expose a whole population to death (1978: 136). This is a biopolitical sovereignty that is 'more powerful' than classic sovereignty because it does not depend on the threat of punishment but by bringing 'life and its mechanisms into the realm of explicit calculations' (Foucault 1978: 143). This politicisation of life ultimately enhances sovereignty and creates the conditions for 'biopolitics' (1978: 141).

Giorgio Agamben has taken up the question of the form of law that corresponds to the biopolitical situation Foucault described (1998, 2005). The next few pages will introduce some salient aspects of Agamben's elaborations that are important for the forthcoming analysis, in particular that life is a politically determined concept and that the authority of law is based on the division between life and death.[7] In the classic form it involved the right to put people to death, while in modern times this also includes life in legal exile from the law, such as stateless persons and refugees. In order to make this argument he takes an obscure figure from ancient Roman law called the *homo sacer* (or 'untouchable man') and connects it to the modern operation of law. In ancient law, the *homo sacer* was a legal category of life that was banned from the legal order. The human law did not apply to the *homo sacer*, meaning it could be killed with impunity; nor was the life of the *homo sacer* included in the religious law, for its death could not count as a religious sacrifice (Agamben 1998: 11–25). It was defined in a negative sense, wholly by the 'unpunishability of its killing and the ban on its sacrifice' (Agamben 1998: 9).

The importance of the notion of the *homo sacer* is its connection to sovereignty and a political structure prior to law (1998: 74). It also prefigures the role of what Agamben calls 'bare life' in the contemporary context. He develops the idea to capture the idea of life that has legal significance only as a living organism, not as a political subject. Effectively, bare life is defined as the necessary outside to the rule of the law. The sphere of discretion to decide when this category applies and what is inside, Agamben argues, is the modern incarnation of sovereignty, which is something both biopolitical and at the heart of the authority of law (Agamben 1998: 164). His critical method is then to ask what our political situation reveals when considered in the light of bare life as part of a political structure prior to law and morality. One of the most evident is what he terms a state of exception (2005), which is a legally defined situation where the force of law suspends the rule of law and begins to operate beyond its scope. No doubt Agamben has in mind the fact that national constitutions contain various forms of provision for declaring a state of legal exception when order is threatened or the nation itself is under attack. In a state of exception, the legal categories that would normally mediate the relation between life and law (ex: rights) are legally suspended. The 'continuing tendency in all of the Western democracies' is to make the exception a normal technique of government (Agamben 2005: 14).[8] Sovereignty, in the legal sense, is the capacity to define the boundary between the normal situation and the exceptional case. Bare life is the form of life that corresponds to the state of exception; it is in an unconditional relation to law, which is the biopolitical essence of sovereignty.

Thus, Agamben's analysis follows Foucault in seeing life as a politically determined concept. However, Agamben emphasises the continuing role of sovereignty in the biopolitical situation that Foucault was describing. The key to that role is the relation between sovereignty and bare life. This issue will be discussed further in Chapter 5, but at this point suffice it to say that Agamben's feeling is that biopower is a correlate to the fact that bare life prior to law is now 'explicitly and immediately political' (1998: 155). This fact actually disperses sovereignty and spreads it around the social body:

> The sovereign sphere is the sphere in which it is permitted to kill without committing homicide and without celebrating a sacrifice, and sacred life – that is, life that may be killed but not sacrificed – is the life that has been captured in this sphere.
>
> (1998: 83)

Defined in these terms, the sphere of sovereignty is internally a terrain of biopower, which, as Foucault demonstrated, involves perpetual processes of normalisation and management of life in all areas. Agamben is particularly attuned to the implications this has for our understanding of political power, because in 'the biopolitical horizon' that characterises modernity, the physician and the scientist move into the 'no-man's land into which at one point the sovereign alone could penetrate' (Agamben 1998: 159). Thus, included as examples of bare life are bodies kept alive by feeding tubes in hospital rooms as well as the inmates at Auschwitz (1998: 164).

Chapter outline

The first chapter draws out key elements of psychoanalytic ethical thought that provide the analytic lens for subsequent discussions (Lacan 1992), which involves de-naturalising the moral role of the good in ethics and instead focusing on the relationship between action and desire. The extent to which the psychoanalytic view of ethics departs from traditional ethical thought is evident in the suspension of moral judgement of desire, a suspension that makes it possible to consider the function of the good as a limit to desire. This is fleshed out in a series of meditations on the Greek tragedy the *Antigone*.

The second chapter uses the Lacanian analytical framework to analyse new natural law theories in the context of the decline in the perceived legitimacy of national sovereignty in international law. It suggests there is a renewal under way in natural law theory that de-emphasises theological concepts. In particular, the chapter analyses the move away from theological conceptions of a divine lawgiver (i.e. God) in favour of first principles found in nature and reason, which are seen as universal and basic to human survival. The chapter argues that the new natural accounts, while manifestly founded on rational principles, retain a theological core.

The third chapter applies the theoretical concepts to the bioethics policy recently implemented in Canada for university research. *The Tri-council Policy Statement: Ethical Conduct for Research Involving Humans* (1998) formalised review procedures for all research involving humans using a bioethics framework that had already been in place for medical and pharmaceutical research. To accommodate social sciences and humanities research, the biological definition of 'life' was broadened to include social and psychological conceptions. The chapter concludes with a demonstration of how the psychoanalytic view of ethics overcomes a key impasse in the critical debates by understanding ethics in relation to desire.

The fourth chapter explores more deeply the bioethical paradigm, specifically the work of Hans Jonas, philosopher and seminal figure in the development of bioethics. Jonas' thoughts on ethics were a response to what he saw as the historically unprecedented intensification of two basic tendencies of the human species: the drive for survival and the growth of power. However, the chapter argues that Jonas' version of ethics is only possibly on a radically revised political ethos. It is a form of jurisprudence that emerges internal to a form of law based on the power to define the limit of death.

The fifth chapter further addresses the current disconnect of bioethics from political concepts in light of the preceding chapter. It draws on theories of biopolitics to argue that 'life' is a politically determined concept, a notion which is then used to assess the bioethical paradigm, particularly in reference to the work of Michel Foucault and Giorgio Agamben. The recognition of the politicisation of life makes it necessary to go beyond the moral force of the idea of life and intuitive acceptability of 'survival' as a natural law and instead contextualise such ideas as limits in a broader political situation.

Chapter 6 also tackles the question of what it means to engage in politics in this context. Some suggest it is necessary to think in terms of 'limited sovereignty' vis-à-vis international law and/or human rights (MacCormick 1999; Ignatieff 2001). However, the chapter suggests that the so-called decline of national sovereignty is not synonymous with the decline of sovereignty *per se*, as is often assumed in international law and human rights. It suggests that the spread of bioethics throughout society is related to the rise of 'biopolitical sovereignty'. Addressed in the conclusion is the future of human rights and democracy, and whether they may be regarded as limits to political thought and action in this context (Žižek 2006).

Chapter 1

Law and ethics

> Is there anyone who doesn't evoke Antigone whenever there is a question of a law that causes conflict in us even though it is acknowledged by the community to be a just law?
>
> (Jacques Lacan 1992: 234)

The goal of this chapter is to introduce the ethics of psychoanalysis and several concepts from Lacanian theory, using Lacan's discussion of the *Antigone* tragedy and selected strands of its critical reception as a point of entry. In particular, it identifies and investigates the ethical dimension of desire that Lacan argues is exemplified in the dramatic action of the play. The chapter then elaborates the implications of the focus on desire for conceptualising ethics in terms of desire. It suggests that the psychoanalytic take on ethics marks a substantial and formal difference from other forms of ethical thought that preceded it in that it can conceptualise ethical action beyond ideals of the good. The meaning of this is explained by addressing criticisms of the approach articulated by Judith Butler (2000a). The chapter contextualises the criticisms and uses them as an occasion to further explore how the Lacanian approach posits an originary and irreducible relationship between law and desire. The discussion concludes by developing the Lacanian approach to ethics into a framework for analysing ideals of the good from a non-moral perspective.

The *Antigone* tragedy

The *Antigone* is the third of Sophocles' Theban plays (1974). It centres on Antigone, one of the four children of Oedipus and his mother and wife Jocasta. She grew up journeying the countryside with her blind father, who was exiled from the city, but Oedipus is dead and she now lives with Creon, her father's brother-in-law and current king of Thebes. Antigone's sister Ismene grew up in Creon's house and continues to dwell there. Their brothers Polyneices and Eteocles are dead; Polyneices killed in a military campaign against the city and Eteocles in defence of the throne.[9] In the aftermath of the civil war led by the Labdacus' sons, Creon decrees that while Eteocles is to be buried with state funeral rites, Polyneices' body will remain exposed on the battlefield and not receive any rites at all. Antigone will not abide this and she violates Creon's edict by burying her brother anyway. Upon the body's discovery Creon reiterates his edict, this time threatening the penalty of death to anyone who disobeys. Antigone is relentless and buries her brother again. When the act is discovered she is accused and does not deny the deed. Creon then condemns her to be sealed alive in a tomb. Despite pleadings from his own son and the soothsayer Tierecias, the king refuses to relent on this decision (Lacan 1992: 263; Butler 2000a: 47). However, ominous signs of divine anger begin to appear. These finally make Creon change his mind but it is too late. The tomb is unsealed and it is discovered that Antigone has already hanged herself. In the end, the houses of both Creon and Oedipus are destroyed. Creon's son and wife commit suicide and Ismene is doomed to live a solitary life, cursed and stigmatised.

The confrontation between Antigone and Creon is characterised as a dramatisation of the conflict between formal law and ethical conscience. The most influential interpretation for modern generations in this regard is probably G.W.F. Hegel's (1977). Hegel saw the tragedy in terms of a conflict between two forms of justice related to different forms of ethics, one associated with Antigone and seen as pre-political and based on *diké* (the unwritten law of the gods), and the other represented by Creon and seen as based on the institution of the rule of law or *nomos* (Douzinas and Warrington 1994: 43). Hegel's interpretation is that Antigone's final act of suicide signals the eventual reconciliation of the contradiction between unwritten and pre-political ethical obligations to clan and kin and the rule of law in a formal legal sense. For Hegel, the tragedy could be seen as a dramatisation of a metaphysical and historical progression that will see the unconditional call of justice sublimated into the legal principles of a state. The conflict is fully resolved with the actualisation of eternal justice, achieved in a state that establishes a universal rule of law. Hegel's account has been extremely influential, and while each of the writers above makes their own contribution, they are all also speaking back in some way to Hegel's idea of the harmonisation of formal legal right and justice.[10]

One important strand of commentary in this regard is informed by Jacques Derrida's critique of the concept of reconciliation itself in Hegelian philosophy (1986). In much of his writing, Derrida's critical targets are the seemingly straightforward categories of political thought, which have achieved the appearance of stability by virtue of Hegel's logic of the dialectic (Kellogg 2003). In Derrida's reading, Antigone does not usher in a universal rule of law but rather stands as a reminder of an eternal and irresolvable conflict between conditioned written law and unconditional justice (Derrida 1986), her tragic fate thus revealing an aporia in the concept of law itself (Fitzpatrick 2000). Instead of reconciliation, Derrida argued that the actual outcome of

Hegel's formula is an 'eternally circulating and irreconcilable antithesis' (Douzinas and Warrington 1994: 30). The reconciliation that Hegel foresaw on the horizon for humanity never arrives; it remains forever 'to come' (Derrida 1992: 28). Justice is only ever a factor in conditioned legal decision-making, which it also exceeds. Derrida shares Hegel's understanding of Antigone as representing something unconditional and 'eternally other' (Douzinas and Warrington 1994: 21).[11] The difference is that Derrida suggest the impossibility is primary and focuses on the way that the appearance of reconciliation depends on hiding Antigone's body, so to speak. In Derrida's hands the split in the 'ethical substance' that Hegel saw remains forever open.

Marking a distinct departure from the Hegelian interpretation, Lacan finds significant the fact that Antigone never actually claims to be motivated by the sacred rights of the dead and the unwritten laws of the gods. Her desire puts her in touch with the force of 'a certain legality' that is not articulated in any law, human or natural (Lacan 1992: 324). At no point does she explicitly argue that she has a duty to family that overrides the concerns of the state. She never makes the claim that she or anyone else has an obligation to kin that takes precedence over obeying the written laws (Lacan 1992: 279). Her action goes beyond any concern for her own; it drives Creon 'beyond all limits' and makes the community 'lose its grip' (Lacan 1992: 268–69). Antigone's desire appears to be a '"crazy", unaccountable event' from the perspective of the community (Žižek 1997: 223).

From this perspective, Antigone's second violation of Creon's law is an illustration of a more radical aspect in the relation between the subject and signification, which is, at least formally, a universal relation (Walsh 1999: 98). Thus, Lacan argues that Antigone's desire to bury her brother is based not on his particularity or the family's historical specificity but on 'no more than the break that the very presence of language inaugurates in man' (Lacan 1992: 279). There is no justification for what she does beyond the fact that Polyneices is her brother. Lacan also notes that unlike all of Creon's speeches that are 'developed with an end in view', Antigone's justification for her action essentially has no point. This is because it happens after the moment when Creon asks her to explain herself; she justifies her action with 'the martyrdom we have witnessed already behind her' (Lacan 1992: 254–55). *Contra* Hegel, Lacan also argues that Antigone is not acting on the basis of an ethics of family when she insists on burying her brother. This is the significance of her statement that she would not have violated Creon's edict for a child or husband but only for her brother: 'it is this alone which motivates me to oppose your edicts' (1992: 279).

In this sense, she does not desire to obey or resist the law and is not motivated by the morality of her actions. Thus, everything would have been all right if the community had been able 'to cover over everything with the same funeral rites' (Lacan 1992: 283), just as everything would have been all right if Antigone had ceded to her desire. This is not what happened, though; unlike her father, Antigone is not the victim of tragic blindness but instead knowingly desires her eventual fate. The consequences of such desire emerge for everyone when she abandons any attachment to her life. It is not her exclusion *per se* that is the issue for Lacan, but this fact of her desire. Ismene grew up in Creon's house: she is willing to abide Creon's laws and urges Antigone to do the same. Antigone wants something else and, in this sense, it is an inherent feature of her desire that pushes her to 'the realization of something that might be called the pure and simple desire for death as such' (1992: 282). Consider Lacan's highlighting of the Chorus as 'the emotional means' of tragedy. He says to his audience that in tragedy 'your emotions are taken charge of by the healthy order displayed on stage. The emotional commentary is done for you' (1992: 252). Lacan suggests that the Chorus exemplifies the function of the pleasure principle.[12] This is why it repeatedly implores Creon to make an exception for Antigone, even though it also says that it regards her as inhuman and monstrous, and fears having someone like her as a neighbour. The Chorus' protest against Creon's edict is not against the death penalty nor is it made in the name of justice, but instead is delivered out of fear of the extreme 'point of apocalypse' to which everyone is heading (Lacan 1992: 207). Her desire draws her and Creon both 'into the realm of death' (1992: 248). In this sense, the problem with Antigone's action for the city is not primarily that it transgresses the law but that it brings the legal order itself to a point of 'radical annihilation' (Lacan 1992: 212). By emphasising this aspect of Antigone's action, Lacan is suggesting the operation of something beyond the good that cannot be absorbed by a legal order. While none of our fates may 'reach the tragic level of Antigone's', we all have a bit of Antigone in us in the sense that we also must negotiate unwilled aspects of desire (Lacan 1992: 300).

One way of unpacking the implications of this formulation is through Judith Butler's critique of Lacanian psychoanalysis, which is based on the claim that Lacan's formulation denies the historically constituted nature of all social structures and forms of power. In particular, she argues that Lacan's theory is abstractly universal and denies the possibility that the heteronormative dimensions of the social world are contingent and open to contestation (Butler 2002). In *Antigone's Claim*, Butler argues that this is exemplified in the fact that Lacan makes the central concern of his interpretation the nature of Antigone's desire and not the nature of the symbolic order in which she is located. Indeed, it is true that Lacan is interested in the unconditional aspect of desire that Antigone displays in burying her brother, which is not just puzzling for the Greek Chorus but also for interpreters who continue to make disparate meanings of it (1992: 207).[13] The next section will explore the salient features of the Lacanian concepts germane to this debate, such as the symbolic order of language and discourse, the unsymbolisable *real*, and, in a more limited sense, the imaginary order of identification. I will then connect these concepts to the details of Butler's critique and to the meaning attributed to Antigone by Lacan.

The limit of the good

According to Lacanian theory, every subject is born into a situation of meaning and signification that is organised through networks of intersubjective relations, which are experienced as mysterious, unknown and inescapable. Lacan calls this realm of signification the symbolic order of language and discourse (2006: 228–29, 276–77). The real in Lacanian theory stands for a kind of quasi-metaphysical ungroundedness of all symbolic constructions, the radical 'lack' of referent upon which all symbolic meaning depends. The fact that humans create shared meaning that 'superimposes the reign of culture over nature' is an existential response to the radical absence of that division in the first place (Lacan 2006: 229/277). Thus, lack is real in the sense of always implying a '"something else" needed to complete or ground the symbolic', such as faith, anxiety or pleasure (Lacan 2006: 228/276). The real is a name for this lack, which is also the condition for the emergence of human sociality and meaning in the first place. Butler's rejoinder to this has been that the concept of an impossible and unknowable real 'is posited as accountable for the contingency in all ideological determinations, but is never subject to the same logic of contingency that it secures' (Butler 1993: 195–96).

In other words, while Lacan might not be referring to a classically oedipal desire for the mother in the way that Freud was, his concept of the real nonetheless contains the same idea of something that is 'always in the same place' relative to the symbolic (Butler 1997). This suggests there is some inevitable subjective pivot point for entry into language (Butler 1997). As she writes:

> On the one hand, we are to accept that 'the Real' means nothing other than the constitutive limit of the subject; yet on the other hand, why is it that any effort to refer to the constitutive limit of the subject in ways that do not use that nomenclature are considered a failure to understand its proper operation? Are we using the categories to understand the phenomena, or marshalling the phenomena to shore up the categories 'in the name of the Father', if you will?
>
> (Butler 2000b: 157)

For example, in 'Is Kinship Always Already Heterosexual?' she suggests that a Lacanian notion of the symbolic has been invoked in France to argue against gay marriage on the basis that heteronormative kinship structures are the necessary symbolic coordinates required for a child to have a healthy social life (Butler 2002).

What is significant in this context is that Antigone's act of defiance occurs and has effects despite the fact that she is excluded from the political community (Butler 2000a: 9). Her form of action is premised on the very impossibility of the universality of law that Hegel imagined (Butler 2000a: 53). The lesson that Antigone teaches relates to the stability of normative formations premised on the fact that some lives are socially unintelligible and excluded because of what they are. In this sense, Butler takes Antigone's very form of life as testimony to the objective impossibility of universal laws or norms, even those 'of the father' (Butler 2000a: 82). Antigone is not a harbinger of justice, nor is she strictly speaking the 'remains' of its impossibility. Thus, the normative conditions that require the exclusion and abjection of some lives should be the focus of attention and their overcoming the key to any politics that seeks to approximate justice (Butler 2004: 90).

The decisive battleground in this regard is the limited definitions of the category 'human' that are operationalised in particular political situations. These moments are where political resistance can work progressively through transgressing the limits of those definitions in order to include in them lives that would otherwise be unintelligible and excluded. In *Precarious Life*, Butler refers to Foucault and theories of biopolitics in order to make this argument; her definition of 'politics' is a matter of deciding the limits of what counts as human life (Butler 2004). In another work she articulates it in the following way:

> What challenge does that excluded and abjected realm produce to a symbolic hegemony that might force a radical rearticulation of what qualifies as bodies that matter, ways of living that count as 'life', lives worth protecting, lives worth saving, lives worth grieving?
>
> (Butler 1993: 16)

In this conception, politics is not a matter of reconciling law and ethics but of negotiating the division between included and excluded forms of life. The hope is that perpetual normative transgressions of such divisions might achieve a critical mass eventually allowing for a 'a new field of the human' to emerge, where all lives, even one like Antigone's, could be included (Butler 2000a: 9). This new field could potentially exceed civic, ethnic, national and cultural divisions and be imbued with a fundamental ethos of inclusion. She writes how such an 'ideal of a radical inclusivity' is 'impossible', but should still 'govern the political field' (Butler 1993: 193). Antigone's claim for the social recognition of her brother's death is exemplary in this sense. The suggestion is that such a politics may correspond to justice in the contemporary context of globalisation and declining national sovereignty. According to Butler:

> [it] is the condition by which a concrete and expansive conception of the human will be articulated, the way in which parochial and implicitly

> racially and religiously bound conceptions of human will be made to yield to a wider conception of how we consider who we are as a global community.
>
> (Butler 2004: 90–91)

Understanding this political agenda informing Butler's work is one way of approaching her critique of Lacanian psychoanalysis, which, as mentioned above, is based on the claim that Lacan's formulation denies the historically constituted nature of all social structures. In particular, she argues that Lacan's theory denies the possibility that the heteronormative dimensions of the social world are contingent and open to contestation (Butler 2002). The insight helps explain at least one thing that critics have found puzzling about Butler's use of Lacanian theory in general:

> What remains inexplicable in Butler's recent work, however, is her continued insistence on taking on Lacanian theory when this engagement amounts to, in effect, a non-engagement. In her many references to Lacan, she repeats the misreading that other critics have pointed to in her earlier work, most notably concerning the category of the real. Perhaps we are to read her adamant refusal to heed the call of (what she has cast as) the Lacanian law as of the same order as that which propels Antigone's obstinacy in the face of Creon's commands.
>
> (Tuhkanen 2003: 144)

In Butler's conception the issue is the way the law reinforces normative divisions between lives considered socially legitimate and those relegated to the limits of social intelligibility. Her interpretation of the *Antigone* is motivated by a political ethos.

Thus, one of the broader implications of recourse to the real in political theory is a lack of attention to the operation of the normative and institutional processes by which only 'certain kinds of lives' are excluded by symbolic categories in the first place (Butler 2000a: 54). This omission she suggests is why Lacan focuses on an 'inherent feature of [Antigone's] desire that leads her inexorably toward death' instead of the contestable limits of the symbolic order (Butler 2000a: 50). Therefore, one of the agendas of Butler's discussion is to demonstrate Lacan's 'blindness' to the fact that it is precisely his concept of the symbolic that is 'challenged by [Antigone's] kind of claim' (2000a: 53). When it comes to the *Antigone*, Butler claims that while Hegel stood for the immutability of the laws of the state, Lacan's interpretation is based on the assumption of an 'intractable law of kinship' (Butler 2000a: 40). In other words, the concept of the real might signal the formal contingency of norms, but by basing that contingency on the existence of a dimension that lies beyond human action, it also ensures that we are 'virtually powerless to alter the law' (Butler 1997: 98). What this amounts to, according to Butler, is a concept (the real) that functions to secure a formally contingent – yet still practically necessary – conception of law (1997: 83–105).

In order to assess the validity of this charge it is important to keep in mind first that Butler's references to Lacan's concepts of the symbolic and the real are drawn primarily from an early seminar, 'The ego in Freud's theory and in the technique of psychoanalysis 1954–55' (Lacan 1988). Therein, the real is indeed conceived as 'the subject's radical limit', a foreclosed, non-symbolisable and untouchable object that both grounds the symbolic order of discourse and language and consolidates subjectivity (Butler 1997: 122). Butler accurately notes that from the perspective of this limit, Lacan describes 'all resistance' to discourse as 'imaginary' (Lacan 1988: 321). The point that I wish to make relates to the context in which Lacan made these statements. The comments were delivered from the perspective of an analyst facing 'imaginary resistance' to the lack in the analytic discourse and the inability of the analyst to give them happiness. In general, the seminar was concerned with recuperating Freud's important insights about the functioning of the ego in the technique (practice) of analysis. In this context, Butler is right when she writes that Lacan links imaginary resistance with perpetual defeat, but this notion of defeat must be put in the context of the clinic. When Lacan states that the imaginary resistance 'determines the interruptions of the fundamental discourse' he is not referring to legal discourses or oedipal relations in a normative or political sense (1988: 324). Psychoanalytically speaking, it is true that 'law' for the individual psyche is in part an imaginary relation to social norms.

Indeed, Lacan spoke of how transgression is generated internal to law as an 'intimate exteriority' (1992: 139). To illustrate this he made reference to the Christian prohibition on lying:

> 'Thou shalt not lie' is the commandment in which the intimate link between desire, in its structuring function with the law, is felt most tangibly. In truth, this commandment exists to make us feel the true function of the law ... 'Thou shalt not lie' is a negative precept ... In 'thou shalt not lie' as law, is included the possibility of the lie as the most fundamental desire.
> (1992: 81–82)

The point is that the desire for transgression is still a form of law; 'the expectation to transgress is built into the Law' (jagodzinski 2004: 66). This is the basis upon which Žižek has characterised the real as what divides our attachment to the law (Žižek 2001: 92). It also indicates that Butler's contention that 'the dual possibility of being both constituted by a law and an effect of resistance to a law marks a departure from the Lacanian framework' (1997: 98) is leaving out part of the picture.

The imaginary attachment that the analyst encounters in the clinic was only the beginning of the question when it came to the authority of law.[14] At no point does Lacan argue that Antigone's action is a case of imaginary resistance. In fact, the significance of Antigone's desire is the way it exceeds this relation. This is one reason why Lacan is interested in tragic action in the first place and why he introduces the ethics of psychoanalysis with references to Aristotle on the cathartic effects of tragedy. Tragedy engenders a 'purgation of the emotions', which Lacan likens to the imaginary attachment to the law (1992: 244). With this in mind we are in a position to assess more fully Butler's criticism, but to do so we must go beyond contextualising Lacan's comments as clinical elaborations and towards an understanding of the overall structure of his ethics.

Beyond the good

One of the most unique elements of Lacan's interpretation of Antigone is his contention that she does not claim to be motivated in defence of the sacred rights of the dead or by unwritten laws of the gods, as many commentaries claim (Keenan 2005: 118). He held that the exchange that occurs between Creon and Antigone after she has admitted to burying her brother for a second time has been incorrectly translated and as a result incorrectly interpreted (Lacan 1992: 279). She is usually translated as saying to Creon that his edict 'did not come from God' (Sophocles 1974: 429). Antigone's motivations are usually seen as an ethics of family and the laws of the gods, but Lacan has a different interpretation. He asks: is it really true 'as the classic interpretation would have it' that she is 'the servant of a sacred order, of respect for living matter … ? She says clearly "you made the laws". But once again the sense is missed' (1992: 278). According to his translation, Antigone actually says, 'For Zeus is by no means the one who proclaimed those things to me', nor is it 'justice' (Lacan 1992: 278). The real reason is revealed later in her penultimate speech:

> If my husband were dead, I might have had another,
> And child from another man, if I lost the first.
> But when father and mother both were hidden in death
> No brother's life would bloom for me again.
>
> (Sophocles 1974: 976–79)

This puts us in a position to understand the significance Lacan attributes to the 'scandal' of Antigone's desire and her expression of regret after the fact (1992: 280). Antigone appeals not to a higher law or notion of the good, but speaks the name 'brother'; her brother is her brother. In other words, Lacan is saying that at no point does she argue that she has a duty to family that overrides the good of the political community. She never makes the claim that she or anyone else has an obligation to kin that takes precedence over obeying laws, be they human or divine (Lacan 1992: 280). Rather than a clash of different ethics and claims to justice, the import of the tragedy lies in the limit that this 'terrible, self-willed victim' illuminates beyond the good (1992: 247).

The idea of a beyond the good can be identified with the concept of the real, albeit in a slightly different way from Butler's characterisation mentioned above. The real figures here as the incomprehensible 'thing' that Antigone's sights appear to be set on, which motivates her and draws her out of the 'circle of everyday feelings and considerations, passions and fears' that are signified by the Chorus (Žižek 1989: 177). It is real in the sense of being absolutely incomprehensible from a point of view internal to the normative order (Copjec 1994: 10). In other words, there is something about her desire that cannot be absorbed by the political and social reality of her situation. We can see what is at stake if we consider the meaning Butler gives to Lacan's reading of the tomb where Antigone kills herself. She writes that from the Lacanian position:

> The symbolic might be understood as a certain kind of tomb that does not precisely extinguish that which nevertheless remains living and trapped within its terms ... On this reading, the symbolic thus captures Antigone, and though she commits suicide in that tomb, there remains a question of whether or not she might signify in a way that exceeds the reach of the symbolic.
>
> (Butler 2000a: 44)

Indeed, the tomb does signify the limit of the symbolic order of discourse and language. Lacan states that the 'sepulchral mound' is 'the limit of the politics of the good, of the good of the community' (1992: 233). He emphasises that in the end 'we don't know what happened in Antigone's tomb' (1992: 269). However, he also describes it as a 'blindness effect' induced by a certain desire for death, which is not extinguished in the tomb but blazes brightest inside. The association between these events and Antigone's action is the excess of 'death that crosses over into the sphere of life' (Lacan 1992: 248). Thus, it is not the point where Antigone's life is captured – she has already described her existence as 'a form of living death' – but the point where the power of the symbolic is suspended.

Lacan suggests that this is why the Chorus is not so much disapproving of Antigone as afraid of her. Her action provokes the law 'to go beyond all limits' and 'lose its grip' on the social order (Lacan 1992: 268–69). Morality turns upside down and 'the just [i.e. Creon] appear unjust'; in the end, 'the Chorus transgress[es] all limits' by rejecting Creon's authority and finally singing a hymn to Dionysus (Lacan 1992: 281). Žižek is extrapolating from this idea when he writes:

> [Antigone's act] occurs as a 'crazy', unaccountable event which, precisely, is not 'willed'. The subject's will is, by definition, split with regard

> to an act: since attraction to and repulsion against the act are inextricably mixed into it, the subject can never fully 'assume' the act.
>
> (Žižek 1997: 223).

This is a crucial difference: while Butler understands Antigone's action in terms of the fact that her form of life is impossible by the norms of her community and she still 'lays claim to rageful agency within the public sphere' (Butler 2000a: 35). However, her kinship is not the condition that makes it possible for her not to dwell in Creon's house. Her sister Ismene, despite being born of the same parents, has little problem abiding by Creon's laws and would prefer to do so however, Antigone wants to do something else and no appeal by her sister will change that.[15] Thus, it is true that she is excluded because of who she is, but she also goes knowingly towards her fate, and in fact seeks it out. It is the coincidence of those two elements that makes Antigone so remarkable.

Butler reads Antigone's action as a case of 'promiscuous obedience' to her father and Creon because she responds to an obligation to kin that is non-normative and uses legal loopholes to claim intelligibility. However, Lacan argues that in the end the ethical position that Antigone occupies is 'motivated by no good' (Lacan 1992: 240). While Butler is interested in 'kinship trouble' and 'perversion of the law', Lacan is interested in what is revealed 'over and beyond the question of family and country, over and beyond the moralising arguments' (Butler 2000a: 62, 68; Lacan 1992: 247). What we see in the image of Antigone is 'nothing besides life conjoined to death' (Lacan 1988: 232). She is at the limit of the community in the course of 'the realisation of something that might be called the pure and simple desire for death as such' (Lacan 1992: 282). Creon is borne along by a passion 'to promote the good of all' but Antigone is fixated on something beyond the good (Lacan 1992: 258). This idea is missing in Butler's reading, which is not unrelated to her interpretive collapse of Lacan's topology into just the symbolic and imaginary orders. The result is the reduction of the possibility of ethico-political action to the kind of 'imaginary misrecognition of the symbolic structure' that Lacan described in the clinic (Žižek 1999: 261).[16] It is in this sense that we can understand Žižek's suggestion that Butler's interpretation is not political enough even though it is aiming to politicise everything by rejecting the radical 'otherness' implied by the concept of the real (1999: 111).

The key is to understand that Lacan is analysing the structure of action: not Antigone the character but Antigone the image, a term in an economy of action that produces certain effects. Lacan focuses on the seemingly non-subjective aspects which appear to be 'unmotivated' (1992: 230). In general, Lacanian psychoanalysis is not primarily concerned with 'individuated agential consciousness' but rather with relations that structure a situation (Aoki 2006: 13). The image of Antigone is a 'stopping place' for desire, which points in the direction of an 'unspeakable field of radical desire' beyond the limit (Lacan 1992: 216–17).[17] Lacan argues that the good function as 'the barrier' to this beyond, which Antigone's desire flashes before us (1992: 218). Her so-called 'living death' is the realisation of this beyond, which is one reason Lacan calls it a 'zone between two deaths'. It has nothing to do with justice from a normative point of view.

Lacan invokes the questions at stake in the Greek tragedy, but does so also in terms of the issue that has long faced Western political philosophy, namely how best to organise society for the common good (Lacan 1992: 207). Hegel proposes the famous solution in his *Philosophy of Right* of the bourgeois state, which 'lays down rules of a human organisation founded on need and reason' (Lacan 1992: 208). This pertains to the harmonisation of the needs of the human animal with the rational capacity of the human mind through a set of universal principles. For Hegel, through the use of reason human beings can overcome this antagonism by establishing constitutionally based legal orders. It is worth quoting Lacan at length on this point:

> *Antigone* is a tragedy at the forefront of our experience as analysts ... It seems to me to be what it was for Hegel, although in a different way, namely the Sophoclean tragedy that is of special significance. According to Hegel there is a conflict of discourses, it being assumed that the discourses of the spoken dialogues embody the fundamental concerns of the play, and that they, moreover, move toward some form of reconciliation. I just wonder what the reconciliation of the end of *Antigone* might be ... Let us not forget that in Sophocles' last play, *Oedipus at Colonus*, Oedipus's final malediction is addressed to his sons; it is the malediction that gives rise to the catastrophic series of dramas to which *Antigone* belongs. *Oedipus at Colonus* ends with Oedipus's last curse, 'Never to have been born were best ... ' How can one talk of reconciliation in connection with a tone like that?
>
> (Lacan 1992: 243, 250)

Lacan credits Marx for being the first to see the 'biased, partial and incomplete character' of this solution (1992: 208). However, Marx also continued to hold out the possibility that a reconciliation was possible. In *Civilization and its Discontents*, on the other hand, Freud suggests that human needs and reason will never be reconciled, and that the two terms are not sufficient (Freud 1989).

In Hegel's account it is not recognition of the essentially irresolvable nature of the conflict between life, law and morality that is lacking, but that of the originary relationship between death, law and desire is also occluded. For Antigone's form of the good is based on the 'death which is precisely there under life' (Lacan 1988: 232). It appears as an excess of life beyond the good and on the wrong side of nature. Life is not a limit but a threshold that she crosses by burying her brother in pursuit of her fateful doom. Lacan argues that the death drive is illustrated by the image of Antigone. The term 'death drive' refers to the unconscious desire to return to a non-subjective state, even if that experience was traumatic or was only imagined, having never occurred in the first place (Lacan 1992: 104). In the ancient world such a drive was located in the force of divine fate and approached the level of the gods in its immutability. The Lacanian position, with its rejection of ideals of the good and theological referents for morality, looks at ideals of the good in terms of their function on desire, which is 'a human – far too human – power' (Lacan 1992: 314). This marks a great distance between the tragic rites of ancient Greece and their interpretation in the form of a modern morality, which is what is explored in the next chapter.

Chapter 2

Law without a lawgiver

> [The law] … must certainly not be taken as a homonym for what may be spoken of elsewhere under the heading of justice. On the contrary, the ambiguity and the trappings that this law adopts by virtue of the fact that it derives its authority from justice is very precisely a point on which our discourse can perhaps give a better sense of where its real resources are.
>
> (Jacques Lacan 2007: 43)

Lacan's statement above is a reference to the possibilities that the psychoanalytic perspective opens for rethinking the relation of law and justice. This chapter pursues these possibilities by using the concepts from the last chapter to analyse the role of nature in new natural law theories in the context of the decline in the perceived legitimacy of sovereignty. It suggests there is a renewal currently under way in natural law theory that de-emphasises theological concepts characteristic of classic natural law theory. In particular, the chapter analyses the move away from theological conceptions of a divine lawgiver (i.e. God) in favour of first principles found in nature and reason, which are seen as universal and basic to human survival. Based on an understanding of human life as 'natural reasoning', these accounts argue that it is the nature of law to seek the common good (Finnis 1980: 405). This chapter provides Lacan's analysis of this naturalised relationship between law and desire when 'the guarantor who provides its warranty is lacking, namely God himself' (Lacan 1992: 194).

Classic natural law theory is often associated with figures like Thomas Aquinas and Hugo Grotius, who both presupposed the existence of God as the giver of a divine 'eternal law' behind the natural law that human beings are able to recognise by virtue of our reasonable natures.[18] Contemporary figures following in the natural law tradition, like Fuller and more recently John Finnis, have moved away from the presupposition of a divine lawgiver in favour of the idea that law has a 'natural purpose' all its own that is not externally or metaphysically given (1980). In these accounts, the force of law appears as a natural exigency grounded deep in human nature, society or simply nature itself (Kainz 2004: xv). So, unlike the classical incarnation, new natural law theory refers not to a force beyond human power like the divine, but to a knowledge that is definable and available to human beings. This allows for the argument that the purpose of law and legal authority comes from the existence of a common morality at the heart of any political community (George 1996: 330).

The modern definition of natural law holds that objective moral norms can be discovered and elaborated through human reason; there is no necessary mention of a connection to a God or transcendent element (Rothbard 1983: 3). What is emphasised instead is the idea of 'self-evident basic values, from which by logical inference rational people can arrive at' (Kainz 2004: 46). Here is an influential definition of this incarnation of natural law put forward by Robert Gordis. The reason I choose it is because it is slightly polemical and highly exemplary of the shift to non-theological conceptions:

> Natural law declares that only that law is legitimate which is in harmony with human nature. Second, natural law believes that human nature is constant through time, not necessarily unchanging, but with sufficient continuity to make possible generalizations regarding its basic traits, its needs and desires, its limitations and potentialities. Third, it regards human nature as being universal in space, modified to be sure by environmental factors, but still sufficiently stable to permit a generalized theory. Finally, and most importantly, it regards human nature not as known, but as knowable, through the canons of scientific investigation and rational thought.
>
> (Gordis 1961: 244)

To succinctly rephrase, in the modern definition there is no connection to a transcendental element like the divine to justify the authority of law, only a connection to the objectivity of nature and the certainty of human knowledge. It is not related to an ineffable limit but to the equation of reason and nature (Kainz 2004: 54). The important difference from traditional versions of natural law becomes more acute on this point. The idea of 'nature' is then fairly specific. In the place of a connection to a divine lawgiver whose power and knowledge was beyond human cognition, there is a principle of life defined as rationally pursued. This is one reason why there has been a turn to more ancient conceptions of natural law. Specifically, Aristotelian notions are often invoked in these discourses.

For example, Murphy takes the idea of natural exigency as the central component of his 'natural law political philosophy' and cites Aristotle when he defines the common good as a state where life in the political community is 'fully flourishing' (Murphy 2006: 3). In place of a divine lawgiver, newer accounts are centred more on conceptions of the 'common good' based on the value of life. The common good – a state of affairs where all members of a political community are fully flourishing – is defined as the moral purpose of law (Fuller 1969: 106). It is helpful to unpack the implications of this shift by returning to Aristotle's ethics and politics, in particular his idea of political community. Aristotle is cited in order to translate the idea of an objective morality found in nature into the language of politics however, the implications of such a translation are profound. Thus, at this point the discussion will elaborate some aspects of Aristotle's ancient natural law approach that are relevant to the contemporary discussion. In particular, I am going to focus on the ancient view of nature. What I aim to show is how it is different from the modern appropriation in a significant way, which is not necessarily obvious.

There are two connected elements of Aristotle's thought that are relevant to understanding ancient natural law; one involves the nature of human life and is most evident at the level of ethics, and the other involves the nature of human community and is evident at the level of politics. I'm going to begin with ethics before moving on to the political implications, but what I hope will become clear is that ethics and politics are not at all disconnected in the ancient conception; and what unites them is their shared relationship to nature. Furthermore, the discussion will demonstrate how the idea of nature is different from the modern incarnation in that, in the ancient conception, nature is distinct from but implies reasonableness. For example, we should recall Aristotle's discussions of ethics and brutishness. Brutishness is natural but not reasonable from the perspective of the moral domain. 'Brutish' pleasures are those things that are not naturally pleasurable because they do not lead to happiness if we apply the golden mean.

Aristotle discusses the idea of the good concept in *The Nicomachean Ethics* (1999). In the most general sense, the good is happiness. Aristotle felt that no matter what we are doing we are ultimately always trying to achieve happiness, even if we do not always seek it out in the most prudent way. Thus, happiness is the highest good in Aristotle's morality. Furthermore, he reasoned that the experience of pleasure is what makes human beings happy; thus, if happiness is a universal good, the guide to happiness is pleasure. In this sense, the path to the good is blazed by pleasure in Aristotle's conception of morality. However, true happiness is not attainable by just any kind of pleasure: it is blazed by rational pleasure. The use of reason in satisfying our desire for pleasure is therefore essential to achieving happiness, according to Aristotle. Living a happy life involves harmonising our natural inclination towards pleasure and our natural ability to reason.

As mentioned in the introduction, the method for doing this is referred to as the golden mean. Living in accordance with the golden mean allows for the satisfaction of desires in a rational and moderated way, which was seen as the key to happiness. Living this way engenders happiness or the 'good life' as Aristotle called it. Thus, it would be accurate to say that ancient morality is based on living according to our nature as rational and pleasure-seeking animals to achieve happiness.

What should be highlighted is the basis for the connection between rationality and nature that the concept of the good implies. Happiness as the good is not based on the reasonable naturalness of pleasure alone. The other referent was the realm of the gods. Rational thought brings the highest pleasure for humans not because it recapitulates a structure of nature but because it imitates the activity of a god (Aristotle 1999: 1778x 7–8, 1144b 3–17). We would only engage in this activity if our natures were free of having to negotiate opposing tendencies and if we did not have to occupy ourselves with the need to sustain our mortal existence. Unlike human life, divine existence is not dynamic and involves no real change, no contingency, no degeneration and no urgency of finitude. The gods are engaged in the 'single and simple pleasure' of philosophical contemplation (Aristotle 1999: 1154b 26). In contrast to human existence, the life of a god is the activity of immobility, which is literally death for us. Thus, while the natural purpose of human life is to achieve happiness like a god, unlike the gods humans can only achieve this purpose if the necessities of natural life are secured, thus freeing up mental faculties to engage in the rational thinking necessary to find the golden mean. This security can only be achieved in the context of association with other humans, also known as community.

At this point I wish to move to discussing Aristotle's conception of politics, for the role of a rational definition of nature and natural purpose also comes up in that context, albeit in a different way. The fundamental principles of Aristotle's political thought are found in a text called *The Politics,* where he famously argues that the formation of political community is a natural process (1962). Human beings as individuals are not self-sufficient in isolation (Aristotle 1962: 1253a10–16, 9–38). Therefore, to survive we seek out others and form social groups for security. Aristotle hypothesises how the first human social groupings came together in different ways for the sake of self-preservation. Basic social relations, such as family, slowly gave rise to households that provided for the 'satisfaction of daily recurrent needs' (Aristotle 1962: 1252b8). As with other animals, therefore, it was the concern for basic survival that first drew humans into proximity.

Human beings are unique among animals that also live in groups, such as bees and herd animals, in that we are meant for a higher degree of association because of our ability to speak and use language. The creation of shared systems of meaning and normative frameworks eventually becomes the foundation for more complex communities and forms of association. Aristotle states, that humans alone have a perception of 'good and bad and just and unjust and other things; partnership in these things is what makes a household and a city' (Aristotle 1962: 1253a9, 15–18). In this sense, human beings are by nature meant for political association. But while households naturally combine into villages, the establishment of a political community is based on something more than the need to secure natural biological existence: it is also based on the idea of a happy life (Aristotle

1962: 1252a1–7). If a political community '*grows* for the sake of mere life', it '*exists* for the sake of a good life' (Aristotle 1962: 1252b9–1253a9). Happiness is an individual good in the context of ethics and a common morality in the context of politics.

The common good is actualised when every citizen in a community is capable of ethics and therefore achieving happiness. The individual orientation to ethics is also the basis for Aristotle's discussion of nature in the context of politics. In this sense, the idea of a division in nature is important because of how morality and the good are largely assumed as the basis for politics. A political community is bound together by this common biological vulnerability; however, here the delimited moral domain discussed in the introduction is inserted into politics in the ancient natural law conception. An impetus of sociability might be survival but the goal of political association is happiness defined as the common good. Happiness was thought to be inclusive of all the individual goods of citizens. It is important to highlight where the connection between happiness and the rule of law comes from, for while it comes from nature, it also depended on another reference.

Everything just described about nature and the common good of happiness notwithstanding, for a political community of that sort to actually come into existence something more is required. A political community also depends on the rule of law (Aristotle 1962: 1292b30). The enactment of the rule of law sets the conditions for achieving happiness by guaranteeing individuals' physical security in an equitable and predictable manner. Thus, while the existence of a common morality is essential to a political community in the ancient natural law understanding, the rule of law is also a necessary component because it sets formal criteria for political offices, particularly the sovereign office. The security afforded by the rule of law is what gives citizens the luxury and freedom to pursue a happy life.

However, and here is another important point, Aristotle discussed how the enactment of a rule of law exceeds the common morality already described and depends on another external element. In order to establish the authority of law, something else is required, and this external referent is the person that Aristotle calls a lawgiver. The first enactment of the rule of law is also the founding of a political community, which requires a catalyst in the form of a lawgiver (Aristotle 1962: 1287a28). The result is a constitution. The figure of the lawgiver in Aristotle's philosophy does not play the same role as God as in the later theological versions of natural law. Nonetheless, his or her function is exceptional and parallels the role of brutishness already discussed in the introduction. Unlike a typical member of humanity, such a person does not need the golden mean to moderate their desire. A lawgiver is someone who is born virtuous and moral by nature. Aristotle describes them as naturally trustworthy and incorruptible as leaders. This is why, for example, there is no need to safeguard them against the abuse of power, and why Aristotle said that they ought to be given unlimited power to govern (1962: 1284b25–34).

Thus, 'lawgivers' are no typical politicians but transcend the limits of human nature and 'become gods' (Aristotle 1999: 1145a23–24). There are 'certain divine causes in those who are truly fortunate' (1999: 1179b22–23). Thus, in the context of politics, a lawgiver stands outside the political community where only 'god and intellect rule', and 'from that place', and – quoting Aristotle again – 'asks the law to rule' (Aristotle 1962: 1287a28). Because of their independent disposition, such a person exceeds the normal limits of human nature and their natural connection to morality gives the law validity. Not unconnectedly, such nature is associated with heroic figures who seemed to be in possession of divine attributes. In fact, Aristotle's only example of what he calls a state of superhuman virtue, the mirror of brutishness, is taken from a passage by Homer describing how Hector gave the impression of being born of the gods instead of mortal humans. For, in the ancient conception, individual morality was defined in relation to the realm of the gods.

The connection between law and nature at the heart of the ancient natural law was not derived from the existence of a common morality in nature, but from the exceptional nature of the lawgiver, which exceeded the political community and came closer to the realm of the gods. Because of the divine aspect, a lawgiver does not need a political community to achieve the good life. Such a person is exceptional because they do not need others or a political community in order to achieve happiness. In this sense, the lawgiver exceeds political community and human morality, and the externality of their nature is a necessary condition for the rule of law, where the focus moves from the individual ethics to the conditions for the common good of a political community.

The role of both the lawgiver and happiness in Aristotle's philosophy is important because alongside the decline of sovereignty as an explanation for the force of law a supposedly Aristotelian idea of nature has emerged, or re-emerged, in natural law theory (Kainz 2004). As already discussed, several thinkers are attempting to develop a natural law theory that does not depend on a metaphysical presupposition or external transcendental principle. A central component of this new natural law is the idea that legal authority to control behaviour destructive to the common good is a rationality grounded in nature. On my reading, however, the move to ancient sources does not accomplish this. There are two reasons for saying this. First of all, it was the realm of the gods that designated happiness as the highest good, which demarcated a moral domain based on a division of desire and pleasure, and there is the role of the lawgiver in constituting a legal order. In ancient natural law, this figure works as an outside supplement to the natural purpose that new natural law theory valorises. Second, 'nature' in the ancient universe is distinct from reasonableness *per se* (unlike divine existence) but living in harmony with nature is most rational for human beings.

Thus, nature is not simply identified with rationality, but with a division between the existence of a divine dimension that gave law its force and delimited the content of morality. This is important in itself; however, I wish to consider its implications beyond the point of critique of new natural law's uptake of Aristotelian conceptions of nature. For the question remains of the limit function that the divine played in defining reasonableness in the ancient natural law. In order to pursue this question, the discussion will return to the *Antigone* tragedy, for in Greek tragedy this divine element appeared in the form of *até*, or fate.

Desire for death

The word *até* is translated as 'fateful doom' (Krell 2005: 13). The word referred to a force sent by the gods that resulted from hubris, the error of pride. In Greek tragedy, *até* is operative in the passion that compels the hero unwittingly towards a destructive and violent end. It was an Oedipus-like blindness to the destructive consequences of one's actions (Lesky *et al.* 1996: 124). The term *até* refers to a sphere beyond which the justice of the gods rules (Lacan 1992: 259). It was the inevitable outcome for those who transgressed into the domain of the gods' transcendent power and was always associated with the threshold between the human and the divine (Altizer 1985: 37).[19] It was retribution, where the natural inclination of the hero comes to work against them and propels them to their own downfall. *Até* is at the threshold of the distinction between the human and the divine and manifested when the limit was crossed. In the *Antigone* tragedy, Antigone's desire for death and her position in relation to that desire are what occasion the manifestation of *até*, because her resulting action challenges the limit of the human and the gods.

However, in Antigone's case *até* is not a limit but the goal of desire. The truly monstrous aspect of Antigone's action is her desire. She is not an unwitting victim, but knowingly pursues her fate. Her desire for the death that lay beyond the tragic fate that it calls forth confuses the Chorus. Thus, Antigone's desire is 'inhuman'; the term the Chorus uses to describe her is 'raw', which literally means uncivilised (Lacan 1992: 263). This response, according to Lacan, shows that the Chorus 'doesn't understand anything' about the desire in question (ibid.). To appreciate this we will consider Niobe, another mythic character that Antigone compares herself to during one of her laments, who also experiences a tragic fate sent by the gods.

Niobe was a goddess who had fourteen children with Amphion, a half-god and co-founder of Thebes. She boasted of her superiority in having more children than Leto, who bore Zeus two sons, Apollo and Artemis. Because she boasted of something that was ordained by the gods – and so was hers by fate – for this hubris the gods killed her children (Cook 1964: 6–30). The gods' response was an example of *até*, a 'manifestation of their existence' which 'establishes a law far more than it punishes for the infringement of one already existing' (Benjamin 1968: 294). The killing of her children is akin to sovereign violence, which reasserts the fact of a limit. The killing of her children is not a matter of justice or morality in a normative sense, but the instantiation of limit that makes normativity possible (Benjamin 1968: 294). After unwittingly calling down the force of the gods, Niobe asks Zeus to turn her into stone. In Greek mythology she figures as a 'boundary stone on the frontier between men and gods' (Benjamin 1968: 295).

As she is being led to her tomb, Antigone likens herself to Niobe 'become petrified' (Lacan 1992: 281). She invokes Niobe to suggest that her experience of life is a living death, like the experience Niobe felt as she changed into stone. The Chorus responds by telling her that it is a great glory from their point of view to share the same fate as a goddess (Sophocles 1974: 149). They say to her that although her mortal life is ending, her name will be 'undying' (ibid.). Antigone's reaction is offence. She is outraged by the Chorus' association of her action with the gods (ibid.). She has a different relationship to *até* than does Niobe because she is not ignorant of its imminent approach. Unlike the conventional tragic hero, Antigone goes wilfully in that direction (Lacan 1992: 277). At this point the Chorus says to her that her act of homage is good in itself but that the authority of law cannot afford to look the other way and collude with the transgressed (Sophocles 1974: 149).

However, consider how Creon's tragic fate is consummated in a more standard sense. He crosses into the sphere of *até* and inadvertently wreaks havoc because of the desire to make Antigone a case for the law's application. Once the Chorus anticipates the consequences that are to come, it casts aside respect for the edicts of the city (Lacan 1992: 281). In his final lines, Creon declares he is nothing, alive without life for having killed his son and his wife, who both commit suicide upon the discovery of Antigone's body (Sophocles 1974: 162). The Chorus agrees and states that the guilt for the tragic outcome is Creon's alone (Sophocles 1974: 160).

This is also why Antigone should not be regarded as a representative of justice or morality, even though the aim of performing funeral rites for her brother is a good. It is why it is clear that she is not acting on behalf of a higher purpose in any straightforward sense. Antigone is a law unto herself.[20] There is nothing about Antigone's action that is concealed to her. She in fact states that she would prefer to meet death sooner rather than later because being dead is what she already knows herself to be (Sophocles 1974: 138). This desire and Antigone's action put her directly in contact with her family *até*. In this sense, her transgression is a violation of Creon's law in a secondary sense: it occurs as a result of a transgression of the division between the humans and the gods, which is why Creon is drawn into the fray in the first place. Antigone's petrification and Creon as lawgiver are equally unshakable, unyielding, fixed positions.

Recall in Aristotle's reasoning that the division between humans and the gods lies in the former being subject to natural contingencies and opposing internal tendencies. The unchanging aspect of a god's existence is akin to death from a mortal point of view. The difference is also the basis for the distinction that Aristotle drew between the moral domain (of vice and virtue) and both superhuman virtue and brutishness, which, in this sense, are the same in their being unnatural natures for humans. Antigone is a victim of Creon's fate and her 'brutish' desire. It is her disposition towards her desire that makes her life feel like living death. Creon's error was in trying to include something that is 'by nature' outside of politics.

This is why the Chorus does not want such a person as Antigone as a companion or neighbour, 'nor does it want to have the same desire' (Lacan 1992: 276).[21] This is not because it finds her a moral abomination, but rather because the Chorus fears her desire. There is something unbearable about its trajectory, which is towards death (Lacan 1992: 222). Consider the hymn delivered to Dionysus by the Chorus at the close of the play. The hymn is not a celebration of a victory over unjust laws, nor does it honour her sacrifice in the name of the laws of the gods, but it is rather a sign that a certain limit has been crossed (Lacan 1992: 268–69). The hymn to Dionysus signals a crossing into the sphere of *até*, the place where 'the will, or better yet, the *diké* of the gods rules', and where the Chorus states that humans do not enter (Lacan 1992: 259). From this perspective, Antigone is one who uses the force of *até*. This is what gives her actions a savage dimension.

In crossing the limit between mortals and gods, Antigone indicates the function of the realm of *até* in the ancient legal order. It is a limit that even the king of Thebes will only briefly cross. One thing beyond human power in tragedy, as the Chorus states, is death (1974: 137). In her relation to *até* she functions as a radical negativity and present absence in the sphere of law. Lacan conceptualises her form of life as 'between two deaths': in Antigone's case it lies between the biological death she meets in the tomb and the social death she already suffers in the community. Paradoxically, her negativity acquires an outward form by virtue of Creon's attempt to make her a case for the law's application. Antigone talks of being already dead, but it is only 'when what she already affirmed herself to be takes on an outward form' that the ultimate destructive effects occur when she enters the tomb and finds the death to match her state of living (Lacan 1992: 306). From the perspective of her desire, this is a 'race already run' (Lacan 1992: 272). Lacan's interpretation is not based on 'individuated agential consciousness' (Aoki 2006: 13). The action is presented through a structure of 'subsidence', which refers to the collapse of a structure around an empty space (Lacan 1992: 265–66). The result is 'the piling up of different layers of the absence of the hero in time' (1992: 265). In the course of the action, various 'levels' of desire are decanted: the Chorus, Creon, Antigone (Lacan 1992: 312).

The absence in question is Antigone's very form of life. It is when the law tries to hold to its own protocols and refuses to make an exception that Antigone's absence comes to exist in the political world. Antigone's form of life is death that 'encroaches on life' and suspends 'everything that has to do with transformation, with the cycle of generation and decay, or with history itself' (Lacan 1992: 285). Her 'inanimate' god-like condition blazes at the limit on which *até* is founded and articulated as such (Lacan 1992: 212). This force is associated with the gods and the limit of law in the Greek universe. In this way she 'participates in a certain legality' which is 'of the order of law, but which is not developed in any signifying chain' (Lacan 1992: 278). Lacan put it in terms of the 'break that the presence of language inaugurates in the life of man' (Lacan 1992: 279). According to this she is 'an illustration of the death instinct' (1992: 281).

For Aristotle, this question is already resolved by the certainty embodied in the lawgiver and the idea of happiness, which are two levels where the divine intervenes through nature to fill the role of the sovereign, so to speak. Thus, the authority of law is not based on a common morality so much as the common morality is based on the existence of the realm of the gods. These factors indicate the form of metaphysical presuppositions underlying ancient natural law. When the question is expanded from individual survival to the survival of political states and nations, the details of Aristotle's concept of nature become more pressing. This was demonstrated by the discussion of the beyond of *até*, a divine and transcendent limit zone internal to the division between human and divine. Faced with Antigone's unnatural desire for *até*, even the king of Thebes is 'excessively limited' (1992: 267).

The limit of life

The reason I have explored the work of Aristotle is because I think that the renewed interest in natural law that draws on his understanding of nature in lieu of sovereignty or theology must be scrutinised closely. It is true that the idea of God as a supreme lawgiver has given way to the notion of the authority of nature defined wholly by reason and scientific knowledge, not metaphysics. Along with this comes the belief that moral principles can be directly known by human beings, a significant contrast with traditional and ancient natural law theory alike. Such a move reinforces the value of Aristotle and the notion of the common good, especially the idea of having a natural impulse to work together politically to attain the common good (Murphy 2006). This is essentially a form of the 'overlap' thesis, for as I have shown it was Aristotle's idea that politics and ethics naturally 'overlap'. However, as we have seen, this overlap requires a divine referent.

The point I wish to make is not that it is necessarily incorrect to view life as a basic value, but rather that in the Aristotelian conception the overlap between law and morality presupposes a definition of nature that is continuous with the divine and, as such, relies on a set of terms that remain external to law and morality. In this sense, humans have a moral end that is in accord with our human existence yet extends beyond it. In other words, the ancient account is not as human-centric as it seems and was dependent on a zone of exclusion that was theological and transcendentally designated. In the contemporary context, natural law discourses in which nature and the divine are (at least explicitly) dissociated have a fundamentally different meaning; the idea of 'nature' is equated with reason and seen as non-metaphysical. Thus, the question to be asked centers on the basis for the equation of nature and reason. Namely, on what basis are the seeming irrationalities of nature excluded?

One of the fundamental principles of new natural law theory is the idea that life itself is a rational good, and that intrinsically the obligation to promote our health and a right to self-defence are legitimate. Difficulties with this idea have to do with the way it defines life as a good in relation to the imperative for survival. To begin to explore this terrain, the next chapter brings the Lacanian perspective to bear on bioethics. It considers recent developments in social science research ethics in Canada, specifically the implementation of bioethical policies for all university research involving humans, as a point of entry into these matters. Bioethics is a form of ethical thought derived from medical ethics that takes 'life' as good, the *primum vivere*. In this regard, it parallels new natural law theory in the tendency to attribute to life the status of an essentially universal consensus.

Chapter 3

Ethics and the good

In 1998 in Canada all university research involving human subjects became governed by a single-based policy. At that time, *The Tri-Council Policy Statement: Ethical Conduct for Research Involving Humans* was implemented, which extended to social sciences and humanities research, a form of ethical oversight that was already in place for medical and pharmaceutical research. The chapter introduces this logic by taking the bioethics application in the social sciences as an initial case of the bioethical relationship between law and ethics. It is an illuminating context insofar as it is on the threshold of the expansion of bioethics into non-medical domains. In order to be applicable, it was necessary to broaden the understanding of 'life' beyond bodily and organic manifestations.

In Canada, *The Tri-Council Policy Statement: Ethical Conduct for Research Involving Humans* established the requirement that all university research involving humans be approved by a Research Ethics Board (REB). It sets a single national standard for all research on humans and REBs are a regulatory mechanism for ensuring that research meets minimum standards of moral and legal responsibility. Similar requirements exist in a more patchwork form in the American context (Brainard 2001; Lindgren 2002; Gordon 2003). The goal of *The Policy* is ethical research conducted in accordance with some formal rules and review procedures that enshrine key moral principles. The applicability of the rules is based on the idea that there are moral principles common to all disciplines, such as dignity, which it is possible to identify insofar as they are common to human beings. This is indicated in the preamble of *The Policy*:

> The Policy seeks to articulate ethical norms that transcend disciplinary boundaries. The fundamental ethical issues and principles in research involving human subjects are common across the social sciences and humanities, the natural sciences and engineering, and the health sciences. They reflect shared fundamental values that are expressed in the duties, rights, and norms of those involved in research.[22]

The notion of moral norms that transcend the disciplinary differences among the medical sciences, social sciences and humanities has been criticised in various ways. Many question to what extent the ethics of research can really be said to be common across disciplines, given the way the goals and methods of research differ dramatically (Cassell 1980; Christakis 1992; Wax 1995; Gordon 2000; Mauthner 2000; Van den Hoonaard 2001, 2002; Plattner 2003; Lederman 2007). Formal informed consent requirements have been criticised for being too rigid for much social science research (Borochowitz 2004; Halse and Honey 2005; Shannon 2007). Lipson (1994) and Nelsen (1994) both point out that social scientists often work with groups who cannot sign consent forms because of illiteracy or ongoing judicial affairs. Questions about what constitutes consent are also being discussed (Adams 1981; Levine 1995; Scheper-Hughes 1995, 2001, 2002; Gerlach 2002; Caplan 2003). Homan comments that truly informed consent is impossible for qualitative and ethnographic projects, because 'researchers cannot share with their first subjects interests and purposes which may crystallise only after some time in the field, nor are they able to anticipate all the consequences' (Homan 1992: 322). These concerns follow from the notion that general principles present the danger of 'subsuming of a particular situation under an abstract ethical rule' (Mattingly 2005: 465).

Some of the criticism is concerned with issues of workability, but commentators are also taking note of the fact that after their implementation ethics review procedures have tended to expand and become more regulated in Canada and the United States (Shea 2000; Oakes 2002; Marshall 2003; Haggerty 2004; Mattingly 2005). Mattingly comments that 'university and hospital institutional review boards [IRBs] scrutinise proposals and, especially, consent forms, with a level of critical attention unheard of … in the mid-1980s' (2005: 454). These issues have led some to question to what extent a bioethics ethos and review procedure is appropriate to social science research at all (Rose 1999; Lock 2001; Scheper-Hughes 2001, 2002; Rapp 2003). As Bosk writes, 'a set of global rules applied to all research … is hard to fathom' (2007: 196).

One important issue that has emerged is the way 'the Canadian research ethics infrastructure' is involved in an '"ethical creep" whereby REBs have unintentionally expanded their mandate to include a host of groups and practices that were undoubtedly not anticipated in the original research ethics formulations' (Haggerty 2004: 392). The phenomenon of 'ethics creep' is problematic because it 'fetters' social scientists' knowledge production (Haggerty 2004: 391). Haggerty suggests that this will occur because 'the more ethical roadblocks are installed for innovative and critical research', the more 'scholars start to follow what they perceive to be the path of least institutional resistance' (2004: 412).

This tension is constitutive of a divide in social science research ethics. Bosk has noted that a disjunction has emerged 'between our working conceptions of ethics and the official definitions' (Bosk 2007: 196). Haggerty also identifies this issue, writing that 'one of the paradoxes of the formal research ethics system is that there is often a distinct but unquestioned rupture between following the rules and conducting ethical research' (2004: 410). Taking this disjunction as point of departure, my discussion taps into the work on the standardisation and regulation of ethics review, but the aim is to move beyond the problem of the incompatibility of general rules with the ethics of particular situations, and towards an analysis of the relationship between desire and the good.[23] Haggerty is correct that ethics creep is likely to continue to homogenise research design, but whether or not this fetters knowledge production is not my interest in the disjunction in research practice. With the following examples, I begin to trace the dynamics of desire this disjunction indicates. My starting point, then, is the issue of desire on the part of researchers in the disjunction between law and ethics in research practice.

Acting ethically

In 'Toward a Vulnerable Ethics of Research Practice', the author criticises formal ethics policy (Mattingly 2005). She writes also of the general awkwardness of asking potential research subjects to sign informed consent documents because 'not one word sounds trustworthy' (2005: 455). Not wanting to be associated with such formal protocols, she tries 'to convey the impression that I am not the person behind this unfeeling language' (2005: 455). The rupture is emerging here between the formal consent document and the researcher behind it. What is interesting is the nature of the desire for recognition from the research subjects; the consent documents do not 'sound' trustworthy, and the researcher wants to give the 'impression' of being trustworthy. However, this desire is complicated by the way the author cites research situations in which she had to break the rules in order to act ethically. The ethical issue this presents is evidenced implicitly in an anecdote.

We are told the story of a subject who developed 'some level of trust' with the researcher. This person did not want to be 'just another research subject' but instead wanted a special relationship: '[Shanelle] often confronted me by offering me data that supported her sense of having a special relationship with me' (2005: 456). In one case, while accompanying her subject to an appointment with a school psychologist and speech therapist:

> Shanelle introduced me to the professional team as her sister … Since I had not received permission from these professionals to audiotape, and in any case had unexpectedly been turned into a family member, I simply sat quietly while Shanelle and the professionals talked together. When the meeting was over, Shanelle motioned me to her car and there, with a big grin, she pulled a tiny tape recorder out of the purse she had taken into the meeting. She showed me her tiny recorder, still recording,

> removed the tape and gave it to me. I grinned back conspiratorially ... Laughing, I thanked her profusely. To this day that tape sits in my closet at home, untranscribed.
>
> (Mattingly 2005: 456–57)

In discussing why she accepted the tape the author argues there was no choice: 'Refusing such gifts is terribly disrespectful, even unthinkable', despite the fact that one would never admit accepting such a 'gift' to an ethics committee (2005: 456). This reveals the disjunction between following the rules and acting ethically, but there is something else at work too, something that was called unthinkable. It is reflected in the way the author discusses trust and the story of the gift. She expressed concern about sounding trustworthy, and it seems that is at stake when she gives Shanelle the impression they are conspirators by accepting her recording even though she cannot use it. The rationale for the lie is that it was simply an ethical necessity. The researcher had no choice but to break the rules to act ethically.

In the example above, we know why the gift was offered: Shanelle wanted to have a special relationship with the researcher. Trust is described as a critical feature of the research relationship and formal consent documents are criticised for not sounding trustworthy. However, the trust that Shanelle had in being a collaborator is proven groundless by the telling of the story. Thus, there was something disingenuous in accepting the gift, but the moment of decision is concealed behind the idea that not accepting the gift would be simply 'unthinkable'. In this case, following the rules is rejected on the basis of the assumption that the researcher knows what is good in the situation. General principles and following the rules would be wrong because the researcher must respond to ethical necessities in particular situations. The difficulty with appealing to necessity as a rationale for ethical action is that it denies the element of decision in regard to the good. Too often there is an implicit slippage between 'the "ethical" and the "necessary"' that evades the question of ethics (Bosk 2007: 195). I would maintain that appeals to necessity are decisions; even the language used covers that over. However, a focus on the rules alone misses the underlying desire for 'acting ethically'. Furthermore, the potential for harm is identified in the external formal rules, which are seen to potentially block ethically necessary action. Such a view privileges the relationship between the researcher and the rules, but, following the ethics of psychoanalysis, we will also consider the relationship between the action and desire.

Bosk notes that it is a 'failure of reflexivity' to appeal to necessity and deny decision; it also allows researchers to maintain 'the purity of our intentions' (2007: 196, 202). He comments that underneath the criticisms lies a form of pleasure: 'acting ethically' is justified 'as not only necessary but actually virtuous' – 'a display of greater respect for our subjects' social world than empty formalism' (2007: 195). With the following example, we see that the necessity to protect from external harm can be a site where problematic assumptions about research and knowledge continue to adhere despite good intentions. With the next example, it is possible to identify a dimension of desire on the part of the researcher implicated in decisions to 'act ethically' instead of 'following the rules'.

In 'Unravelling Ethics: Illuminating the Moral Dilemmas of Research Ethics', the authors relate 'empirical data and concrete documentation of the practice and implications of ethical decision-making in human research' (Halse and Honey 2005: 2141). They intend to address 'sticky moral questions' that emerge when researchers who 'desire to be ethical and moral' must 'transform a research proposal into an application for ethics committee approval' (2005: 2141). One of these questions revolves around the review board's requirement of defining the research population. The researchers' dilemma centres on how to define the population in a way that satisfies the board's insistence for a 'definitive, homogeneous' definition, but without 'erasing the singularity of research participants' (2005: 2151). The authors are critical of the assumption behind the requirement: 'the transparency of a social and psychological reality' (2005: 2146). The issue of population definition is whether to define it in a way that 'privileges girls' standpoints by limiting the study to girls who self identified as anorexic', which would exclude 'girls who did not accept their diagnoses'; or to 'limit the study to girls diagnosed as anorexic', which would privilege the authority of medical experts (2005: 2144). While I agree that it is important to be critical of such assumptions, I also wonder how singularity is actually 'erased' when a researcher defines a research population for an ethics board.

Perhaps it erases a researcher's singular research design, but for the authors the issue has to do with erasing the singularity of research subjects. Somehow, this happens because a definition of the population limits its potential members: 'The definition of the research population offered to our ethics committees satisfied their expectations and requirements, but it narrowed the research focus by excluding a range of potential participants' (2005: 2151). The review board's demand presents a problem of who defines the research population, but the issue was really the way the study's aim was altered by the necessity of 'following the rules'. As with the case of the gift, the authors here regard the rules as an externality that threatens the ethicality of research. The 'harm' that the requirement to define the research population produced was the alteration of the 'research focus and original aim of illuminating the full range of experience among all "anorexic" girls' (2005: 2144). The desire for the 'full range' of experience is fettered by ethics policy, and in that sense limits the availability of experiences available to researchers, which has implications for the subsequent knowledge produced. At stake is the way that ethics policy 'explicitly subordinates researchers to the authority of research committees' (2005: 2151). Their story seems to confirm Haggerty's fear.

The 'Tri-Council' in the title of *The Policy* refers to the three federal funding bodies for research in Canada (medical and health, science and engineering, and humanities and social science). In 1998 they jointly adopted a policy that was based on what was already in place for biomedical research.[24] Article 2.1(a) of *The Policy* describes 'the requirement in both ethics and law: to protect and promote human dignity'. Under the medical model, the living body is the standard against which harms are measured. The biomedical ethos takes the living body as the good, and harms are measured relative to the health of the living body in the face of invasive knowledge-gathering procedures (Oakes 2002: 448; Cascais 2003). Harm is inflicted by interventions of invasive surgical or pharmaceutical means. Considering that social science is often interested in interpretative questions of culture, experience, relationships, meaning, etc., it is significant that bioethical protocols were originally written by and for biomedical researchers with the goal of protecting subjects from risks of biomedical experiments (Oakes 2002: 447).[25]

The biomedical model binds researchers to an ethical standard. Under the formalisation of ethics review, 'Doctor, do no harm' has its counterpart for social science research in the moral law of non-malfeasance. To accommodate social science and humanities research, *The Policy* broadened the notion of human dignity to apply beyond medical contexts:

> The welfare and integrity of the individual remain paramount in human research … The cardinal principle of modern research ethics, as discussed above, is respect for human dignity. This principle aspires to protect the multiple and interdependent interests of the person – *from bodily to psychological to cultural integrity* … Research involving human subjects is premised on a fundamental *moral commitment* to *advancing* human welfare.
>
> (Sections A and B from the introduction, emphasis added)

The principle of 'human dignity' now also involves guarding the 'psychological and cultural' aspects of health of research subjects. *The Policy* also appeals to an understanding of Immanuel Kant's concept of the moral law, which acknowledges that it is impossible for humans to achieve perfect goodness (the *Summum Bonum*), but that ethical conduct is the constant striving towards this ideal. An illustrative comparison is Kant's discussion of the 'immortality postulate' in his *Critique of Practical Reason*:

> Complete fitness of the will to the moral law is holiness, which is perfection of which no rational being in the world of sense is at any time capable. But since it is required as practically necessary, it can be found only in an endless progress to that complete fitness … The infinite progress is possible, however, only under the presupposition of an infinitely

> enduring existence and personality of the same rational being; this is called the immortality of the soul.
>
> (Kant 1993: 126)

Kant is suggesting that insofar as humans are mortal and finite, we are categorically incapable of pure goodness. We will always have further to go to become adequate to the moral law that demands the highest good. Under the postulate of the immortality of the soul, the finite human life uses the empty form of the moral law to act *as if* it will not die too soon relative to the 'highest good' beyond its reach. This understanding of ethics is based on the principle that human knowledge encounters an absolute limit that is not operative for God.

In Kantian terms, human beings cannot have knowledge of things like the highest good, the meaning of life or the nature of the divine. Knowledge of such 'universals' is impossible for humans because of our finitude. The human faculties of intuition and sensation are properties of organic bodies, which mediate our cognition of the world. This view assumes that things-in-themselves are always inaccessible to human reason, and our wisdom and knowledge are thereby limited by the limits that embodiment imposes on reason. This is why Lacan said in *The Ethics of Psychoanalysis* that no human being can put the moral law into practice: 'Kant himself did not believe it possible' (Lacan 1992: 77). Ethics involves forgoing external or 'pathological' motivation for action, in favour of an unattainable object – 'the highest good' – which we can never achieve (Lacan 1992: 76). Our capacities for experience and cognition are the necessary conditions for knowledge, and these are structured *a priori* through 'forms of intuition and categories of understanding' such as time and space. Thus, the condition for knowledge is also a limit to ethics; because we are finite organisms we cannot know the good directly in itself, nor can we be certain we have actualised it in our action. From this perspective, it is possible to imagine *The Policy* as a programme for infinite progression towards an ideal that infinitely withdraws.

A fair question to ask of Kant's ethics is: why would we choose to strive for ethical action under such conditions? Kant's reply would be that we choose to act ethically because human beings *want* to act ethically – whether we want to want this or not. Kant argued that despite the limitation described above, human reason is still compelled by the transcendent beyond and strives for universality and the highest good. By nature, human reason, Kant writes, seeks to complete the 'regress in the series' of conditioned, empirical judgements seeking to arrive at an 'empirically unconditioned' ground (Kant 1982: 175/584). For Kant, the answer to the desire for what is lacking was to see the desire as the force of law, the very emptiness of which, from the human point of view, implied existence of the highest good beyond mortal cognition and known only to God. Through rational reflection, human reason can give rise to ideals that approximate the highest good to use as a guide for action in different contexts. Kant calls these 'ideas of reason', examples of which include the immortality of the soul, the highest good, the existence of God and the principle of natural purpose. These are not objects of human knowledge but speculative principles for use where we do have knowledge (Kant 1982: 173/583; 1987: 255/375).

This interpretation of Kantian ethics that I have just laid out, in which the moral law is an empty form of law, is the context in which we should understand Lacan's suggestion that the Marquis de Sade's world view qualifies as Kantian in a sense: 'the Sadean world' – even if it is its inversion, its caricature – 'is one of the possible forms of the world governed by a radical ethics, by the Kantian ethics as elaborated in 1788' (1992: 79). On Lacan's reading, the Kantian is similar to the Sadean, whose desire to torture is limited by the mortal body but sustained by a fantasy structured around an ideal and transcendent body. Although impossible, if the body was immortal the torture sessions could go on endlessly (Zupancic 2000: 80). Complete satisfaction is approached only in death, while desire extends beyond. In order to explain this, the next section connects Kant's moral law to Lacan's concept of desire. I do this with reference to a Lacanian reading of one of Kant's less referenced texts, *Religion within the Boundaries of Mere Reason*, by Joan Copjec (1996) who argues that therein Kant presents a different mode of ethical action rather than infinite progress of the finite being, one based on a subjective element that exceeds the autonomous will. I begin with Kant's concept of radical evil and then Lacan's concept of desire.

The moral law

The first reading I presented of Kant's ethical philosophy is the one that was assumed in *The Policy*. I want to put that reading in the context of a distinction that Kant makes between so-called 'legal' actions, which are actions done *in order to accord* with the moral law (doing something in order to be technically 'good' – such as having research subjects sign formal consent forms because they are an REB requirement) and 'moral' actions, which one does only *because* they accord with the moral law (having a passion for the moral law to determine action, as opposed to a calculation of practical benefit). Here is Kant on this difference between a 'legal' and 'moral' relation to ethical duty:

> That a human being should become not merely *legally* good, but *morally* good (pleasing to God) i.e. virtuous according to the intelligible character (*virtus noumenon*), and thus in need of no other incentive to recognize a duty except the representation of duty itself' – that, so long as the foundation of the maxims of the human being remain impure, cannot be effected through gradual *reform* but must rather be effected

> through a *revolution* in the disposition of the human being … And so a 'new man' can come about only through a kind of rebirth, as it were a new creation … and a change of heart.
>
> (Kant 1998: 67–68)

The potential for a subjective 'revolution' is also connected to the reason why we can never achieve perfect goodness; the fact that we even have the capacity for the difference indicates the impossibility of adequacy to the moral law. In this passage, the difference between being legally good and morally good is ontological insofar as it is a different relationship between desire and the moral law.

To be free, the determining ground for ethical action cannot be 'any incentive of nature', but admitted that once we exclude this worldly incentive, 'no determining ground of the free power of choice ought to, or can, be adduced [without being] endlessly referred back in the series of subjective determining grounds' (Kant 1998: 47 fn). This means that 'outer experience … does not disclose the inwardness of our disposition' (good or bad) towards the moral law (1998: 81–82). We cannot know for certain the motivations for actions; we do not know if someone else is morally or legally good. Moreover, we cannot know for sure from inner experience that our call of conscience is really ethical. Our moral intuition when breaking the rules to 'act ethically' is not a reliable measure of goodness. As Kant put it:

> Even a human being's inner experience of himself does not allow him so to fathom the depths of his heart as to be able to attain, through self-observation, an entirely reliable cognition of the basis of the maxims which he professes, and of their purity and stability.
>
> (1998: 82)

Such knowledge is 'noumenal' in Kantian terms – 'directly available only to God' – but humans know that there is a moral law because we feel its force indirectly as guilt, an empty call of duty, a force that Kant thought all human beings shared. Because of the limit to knowledge based on inner and outer experience, all human actions are 'every time (not generally but at each instant) defective' because they lack what is available to the divine: the perfect good (1998: 84). This is what Kant means by describing the human subject as radically evil. Human beings are 'evil by nature' because we are limited in our knowledge of the good; this 'radical innate evil in human nature' is there 'because of freedom' (1998: 56). It would be possible to know goodness through human reason if only 'the supreme subjective ground was not in all cases somehow entwined with humanity itself' (Kant 1998: 46). This is the 'ground qua human' of freedom and ethical action (1998: 47).

We are responsible for assuming our duty to the moral law in how we decide on the principles of our actions, and because it is in theory possible to raise a 'bad' maxim to the point of moral law (as, perhaps, Sade did), all 'maxims must be viewed as accidental' (1998: 46). Kant explains that we can think of humanity as a mode of 'self-love' (1998: 50). Every ethical action happens 'only in comparison with others' (Kant 1998: 51). From this subjective disposition emerges the rational desire to compare equally to others, but also an irrational desire to be the object of others' desire. Kant writes:

> Out of this self-love originates the inclination *to gain worth in the opinion of others*, originally, of course, merely equal worth: not allowing anyone superiority over oneself, bound up with constant anxiety that others might be striving for ascendancy; but from this arises gradually an *unjust desire* to acquire superiority for oneself over others.
>
> (1998: 51)

In this dynamic, the division of radical evil naturally leads to a desire for identity, but it is entwined with aggression and anxiety that others have a 'hateful superiority over us' (Kant 1998: 51). This is where 'jealousy and rivalry' emerge in human relations, and why upon them can be 'grafted the greatest vices of secret or open hostility to all whom we consider alien to us' (Kant 1998: 51). Kant claims this ambivalent disposition 'applies to the species' and is the 'predisposition' to developing a personality (1998: 50). In another idiom, Lacan will describe it as a fundamental dynamic of imaginary relations, for the subject of desire.

One way to begin to think about radical evil in terms of desire is Lacan's conception of the relation between human 'need' and the human 'demand' for satisfaction. 'Needs' are the necessities of life and demand is need meaningfully understood and incorporated into the symbolic order of language and discourse. Even for an infant, need is always already 'inserted in a synchronic world of cries organised in a symbolic system' (Evans 1996: 34). In *The Four Fundamental Concepts of Psychoanalysis* Lacan (1998a) describes how the phenomenon of desire emerges from this:

> Desire is situated in dependence on demand – which, by being articulated in signifiers, leaves a metonymic remainder that runs under it, an element that is not indeterminate, which is a condition both absolute and unapprehensible, an element necessarily lacking, unsatisfied, impossible, misconstrued, an element that is called desire.
>
> (Lacan 1998a: 154)

Demand is divided in the demand for satisfaction and a demand for complete and unconditional satisfaction, which no particular object can ever satisfy: 'the needs which demand articulates may be satisfied, while the craving for love is unconditional and insatiable' (Evans 1996: 35). From the

Lacanian view, insofar as demand divides its object, it also performs a sacrifice of particularity and has a negative capacity:

> Demand annuls the particularity of everything that can be granted, by transmuting it into a proof of love, and the very satisfactions demand obtains from need are debased to the point of being no more than the crushing brought on by the demand for love ... Desire is neither the appetite for satisfaction, nor the demand for love, but the difference that results from the subtraction of the first from the second.
>
> (Lacan 2006: 579/691)

For Kant this desire leads to the ideas of reason and the desire for 'hateful superiority', but for Lacan the question is one of function: 'the particularity thus abolished reappears beyond demand' (Lacan 2006: 579/691). There is an element of desire that continues to seek 'the radical form of the gift of what the other does not have' (Lacan 2006: 579/691). This gift is impossible and unconditional objects do not exist. The desire that relates to them 'bears on something other than the satisfactions it calls for' (Lacan 2006: 579/691). Desire is what preserves 'the structure concealed in the *unconditionality* of the demand for love' (Lacan 2006: 579/691, emphasis added). Desire is related to the demand for an existing object, but also to the form of that demand, which partially negates what exists in order to preserve the fantasy of the unconditional satisfaction of desire beyond its relation to demand. Through a Lacanian reading of Kant's *Religion within the Boundaries of Mere Reason*, this point can be extended to argue that the disjunction between 'following the rules' and 'acting ethically' is not distinct. The force of the rules is indivisible from transgression; the division has implication for ethics, which tries to ground action in the principles of goodness.

Law and desire

This question needs to be addressed in the context of bioethics, which uses medical analogies as a grammar for ethics. Such a grammar 'masks fundamental ethical questions involved in framing justice around images of health, wholeness and collective integrity' (Pavlich 2005: 41). Medical analogies are not appropriate in non-medical situations because they turn truly 'irrevocable moral dilemmas into technically resolvable problems' (Pavlich 2005: 42). In the criticisms of formal ethics policy and procedure described above, the idea of 'particularity' was invoked as natural ground for ethics and general principles as a source of harm. Even when 'acting ethically' the threat to ethically necessary action is assumed as external to the research relation located in the value of life. In so doing, both misrecognise an internal intractable division for something external that can be guarded against. On this point Kant would have agreed with Lacan that the human subject is so constituted that 'there is an ego in him that is foreign to him' (Lacan 1993: 93). This is why both 'inner and outer experience' are insufficient ground for the certainty of moral knowledge. Kant described the relation of self-love, which experiences itself only in relation to others, through others' desire, as the substrate of ethics. While Kant wrote of the necessity of referring to regulative ideals in the course of practical action, he also warned of when these ideals are seen as things 'empirically given' (Kant 1982: 152/506). In other words, what are the implications for ethics if it may no longer be the case, as it was for Kant, that there is something unfathomable about 'humanity'? It is to these deeper issues that the next chapter turns.

Chapter 4

Goodbye to Kant

Bioethics is rapidly becoming a dominant ethical paradigm in academic and institutional settings. It emerged in the mid-twentieth century from within biomedicine as medical practitioners, theologians, philosophers and lawyers confronted and analysed the ethical implications of new technologies and research agendas, like organ transplantation, artificial respiration and gene therapy (Wolpe 2003: vii). What this work shared was an underlying conviction that new medical technologies held the possibility of transforming moral order in ways for which society was unprepared (Stevens 2000: xi). Originally a medical ethics concerned with the health of the living body and patient autonomy, it now encompasses a widening range of domains outside of what would appear to be a medical context. The bioethical emphasis on life is now seen as compatible with the life sciences, biological sciences, engineering, environmental and social sciences. Described as a 'moral science', bioethics combines 'diverse humanistic knowledge' with biological science in order to determine priorities for the 'acceptable survival' of the human species (Potter 1988: 43–45). Furthermore, through its rapid institutionalisation, bioethics has achieved historically unprecedented legitimacy as a form of ethical thought (Stevens 2000: 2).

It is not primarily thought of as a philosophical or speculative ethics, but rather as a practical ethics concerned with how good outcomes can be achieved in particular situations (Allmark 2006: 69). Described as a 'moral science', bioethics combines 'diverse humanistic knowledge' with biological science in order to determine priorities for the 'acceptable survival' of the human species (Potter 1988: 43–45). The emergence of new areas of medical, scientific and technological developments has led to the creation of subfields in bioethics, such as genethics (which arose in response to the Human Genome Project), nanoethics and synthetic bioethics, which is arising in response to the new field of synthetic biology. Synthetic biology aims to create new forms of life that will produce beneficial products for humans, such as medical therapies and biofuels (Parens *et al.* 2008). In this regard, it is significant to note that because of the *ad hoc* quality of bioethical elaborations there is a lack of unification among bioethicists, something that has been described as the 'Balkanisation' of bioethics (Kagawa 2009: 11).[26]

When larger philosophical issues are considered, the discussion typically focuses on the conflicts between the Kantian and Aristotelian images of ethics (Toumlin 2007). On the more formal side (sometimes called 'principlism') there is more of an emphasis on general principles that guide practical action in a given instance, while more particularist approaches focus on the moral intuitions that emerge out of particular events and processes. From this perspective, general theories are above particular interests and recourse to them is appropriate when the solutions to ethical problems are not clear to those involved. More practice-led perspectives often devalue general theories and favour meanings that surface to individuals in particular cases (Allmark 2006: 72–73). They argue that it is preferable to begin with particular cases, not general theories, and then apply the results to other similar cases, until patterns, guidelines, and eventually rules become apparent (Summer and Boyle 1996: 4).

However, some argue that without recourse to general theories cultural differences can lead to irreconcilable impasses (Hare 1975: 203).[27] Formalist approaches express concern about practice-led approaches, pointing out that it will happen that different people will have different intuitions about what acting ethically means in a particular case (O'Neill 2002). Furthermore, they argue that even if consensus is achieved in a given context there is no way of ascertaining whether it reflects a morally correct course of action, or mere skills of persuasion and the force of authority (Smith 1999). On the other hand, concern has been expressed that the use of general theories is often insufficiently attentive to the unique circumstances of particular cases (Allmark 2006).

The more theory-led perspectives are willing to apply seemingly counter-intuitive solutions to ethical problems if consistent with general principles; however, since bioethics is concerned with matters of life and the quality of life, it cannot be said to be successful if its answers to ethical questions are unpalatable to those who have something at stake in particular cases. The formalist/particularist division can be paralleled to the images of ethics already described; it underlies the debates about social science research ethics addressed in the preceding chapter. That discussion demonstrated that the top-down and bottom-up perspectives share an ideal of life as the common good, which manifests as an interest in an individual's life and an interest in the sanctity of life itself.

This chapter will explore the connection underlying the conflict more deeply, for despite the lack of consensus among practitioners of different bioethical approaches, there is nonetheless a certain unity towards the bioethical paradigm, with significant political implications. The political dimensions will become clearer once we start to consider how bioethics might be understood as an overarching paradigm. Because of the way that life functions as an ideal of the good that is neither metaphysical nor defined by ineffability in nature, bioethics appropriates and displaces both the Aristotelian notion of nature and the Kantian notion of the moral law.

The limit of the divine

Philosophically speaking, so-called 'first principles' are primary features about the world that explain phenomena but cannot themselves be explained. They are foundational propositions that are justified only on their own terms and cannot be deduced from anything else. Put differently, first principles are the basis for proving statements and propositions but cannot be proven themselves. An example of first principles in Aristotle's philosophy is non-contradiction, or the principle of consistency as articulated in the *Metaphysics* (Nussbaum 1986: 257–58). Rationalist René Descartes used a method of doubt to systematically arrive at ostensibly indubitable truths. His most famous first principle is *cogito ergo sum* (I think, therefore I am).[28] Kant greatly circumscribed the role of self-evidence in his philosophy and argued that the structure of the mind is the condition under which experience and knowledge are possible. These are the conditions of the possibility of experience, such as time and space, through which isolated perceptions are related into consciousness, thus providing a ground for knowledge (Kitcher 1996: xxviii). Objects that are not amenable to this structure are not objects of human knowledge or experience (i.e. the highest good). Therefore, in seeking first principles of bioethics, the inquiry moves beyond the generalist/particularist debate discussed in the previous chapter.

It builds on that discussion by considering the work of one of the most influential thinkers in bioethics, Hans Jonas. Jonas was a student at the first bioethics institute in the United States (the Hastings Institute) and was involved in debates in the course of the development of the concept of brain death during the 1960s (Stevens 2000: 98). He is recognised as one of the philosophical founders of bioethics. Jonas' work is useful in this respect because of his pioneering responses to ethical questions and engagement with the philosophical ethics of Immanuel Kant. However, an exhaustive account of Jonas' impressive philosophy will not be provided.[29] Rather, selected elements of his work are drawn upon in order to appreciate the broader social and political context in which he defined his interventions.

One of Jonas' most significant contributions was to suggest that the meaning of life has changed. The bioethics definition of life is an Aristotelian-inspired notion of natural purpose. The purpose of life is life; 'life' is defined as something that 'is occupied with the proper care and maintenance of itself and its health' (Dean 2004: 172). The activity of life is active engagement with the environment in order to survive. In this context, life is drawn to itself because it is forever in need of other forms of life to survive (Vogel 1996). Jonas takes the biological concept of 'metabolism' and gives it a philosophical meaning, arguing that a non-willed process is the ontology of freedom. Bioethics is a moral science to the extent that it naturalises an auto-affirmative process of 'having-to-be' against the possibility of non-being apparently found in nature (Jonas 2001: 157–75). This is a universal conception to the extent that it is based on an imperative to live, which 'is ultimately grounded neither in the authority of the self, nor the needs of the community, but in an objective assignment by the nature of things' (Jonas 2001: 283). In a quintessential Kantian form of reasoning, Jonas' thinking is that if human beings have this knowledge, then the conditions for it must be satisfied. The question is what these conditions are.

The moral awareness of which he speaks is based on the existence of a historically new level of power tied to advancements in science and technology, especially biomedicine, which have extended life beyond previous limits, as well as humanity's unprecedented capacity for destruction, which developed in the twentieth century and manifested in the horrors of modern warfare (Hogue 2007). Jonas writes that there is 'no precedent in the standards and canons of traditional ethics' for contending with 'the qualitatively novel nature' of these powers (Jonas 1974: 4).[30] His response is to rethink several of Kant's most important concepts, including the moral law and the highest good, in light of these historical developments.

We recall that in Kant the moral law is empty; humans do not know exactly what it is telling us even though we feel its force in all situations (Kant 1964: 88). The moral law, Kant wrote, tells us in every situation 'to act according to the principle through which you can at the same time will that it should become a universal law' (1964: 421). The form of law posits the highest good as an unconditional universal object, which does not exist in a form that can be directly apprehended by the categories of understanding and intuition that make human knowledge and experience possible. Because of this, Kant is sometimes called a formalist, and his ethics may be regarded as formal in this way (Saurette 2005: 86). The content of moral law is the 'highest good', a thing beyond human knowledge that can only be approximated through the moral law. The form of law relates to the idea that human experience and knowledge is limited in some sense. Only God knows the highest good.

From Jonas' perspective, the growth of human power has rendered what was beyond human limits for Kant – 'an infinite existence of the human species' – into an 'existing' good that is also the highest good (Jonas quoted in Wolpe 2003: vii).[31] This is the point where Jonas departs dramatically from the Kantian position by fusing science and moral values. Jonas overcomes the Kantian division by taking metabolism 'from a heuristic principle to a determinative property of life' (Nichols 2007: 31).[32] In this sense, the first principle in bioethics is also the object of bioethics: life. As a result, where ethics has always been seen as a philosophical concern with metaphysical properties, bioethics is more of a moral science that specialises itself in relation to particular subfields of research involving life (Kagawa 2009: 11).

For bioethics the growth of human power has rendered the Kantian limitation inoperative and the condition of the possibility of ethics is no longer the universal division in the human condition, but the universally shared vulnerability to death. In this sense, bioethics takes up several concepts from philosophical ethics but the deployment must be distinguished from the Aristotelian and Kantian versions, which, although different in many ways, still retained a degree of indeterminacy. For example, Jonas argued that Kant's famous intention set out in the *Critique of Pure Reason* (1982) 'to limit knowledge, in order to make room for faith' can no longer be maintained in the face of what amounts to humanity's metaphysical self-demotion within nature. Jonas goes beyond Kant by equating the 'existing good' of life with the form of the good itself and inserting it into the moral law: 'act so that the effects of your action are compatible with the permanence of genuine human life' (Jonas 1984: 11). In this sense, there is nothing unfathomable about life and what is a moral issue is controlling death. The assertion that 'life knows life' is the other side of the fact that human beings are nature 'become aware of its goodness, and through this awareness, its custodian' (Hogue 2007: 84). Traditional ethical concepts up to the present had been concerned with only the 'short arm of human power' (Jonas 1974: 8; 1984: 6). However, technological, military and medical developments of our times have 'opened up a whole new dimension of ethical relevance' (Jonas cited in Hogue 2007: 83).

While this may seem an obscure philosophical point, it actually has profound political implication, for the ethical conditions Jonas elaborates correspond to broader social relations of power. He is responding to a situation fundamentally different from the ancient universe; the nature and meaning of life is not a divine matter but a determination of human power. Jonas took the role of global guardian to be an ontological grounding for morality and politics. The old form of state-based sovereign law 'dies' to the extent that it loses the pretence to a higher purpose (e.g. the good, justice), and the entire role of law changes 'from bringer of consummation to *preventer* of disaster' (i.e. death) (Jonas 1984: 144). The Kantian limitation has become an abdication of responsibility: we have no 'existential right to ignorance' and must accept responsibility for our 'new role as creator' of life (Jonas cited in Schirmacher 1987: 212). Jonas' personal political position was that formal democracy was only capable of hoping to cultivate the self-restraint necessary for guarding the fate of humanity in the context of this power; however, a socialist system could conceivably enforce the restraint (Hogue 2007).

At this point it is useful to recall that biopower's connection to juridical power was a novel development related in part to technological, scientific and medical advancements since the eighteenth century (Foucault 1978: 137). This resonates with Jonas' position that humanity's capacity for self-destruction is a significant historical development that implies a new form of ethics.

However, Foucault took a different perspective on this capacity as a historical development by not adopting a moral position with regards to it:

> I don't mean that mass slaughters are the effect, the result, the logical consequence of [this] rationality, nor do I mean that the state has the obligation of taking care of individuals since it has the right to kill millions of people … It is this rationality, and the death and life game which takes place in it, that I'd like to investigate from a historical point of view.
>
> (Foucault cited in Dean 1994: 181)

This is an important difference between Jonas and Foucault; Foucault's attention was directed to analysing how the power that correlates to what Jonas regards as a metaphysical condition operates in the realm of politics. In *The History of Sexuality*, Foucault characterised the backdrop for this idea in terms of a shift in political power:

> What might be called a society's 'threshold of modernity' has been reached when the life of the species is wagered on its own political strategies. For millennia, man remained what he was for Aristotle: a living animal with the additional capacity for a political existence; modern man is an animal whose politics places his existence as a living being into question.
>
> (1978: 143)

From Foucault's perspective, it opens up a new dimension of politics that can guarantee an individual's continued existence, but only because of a monopoly on the power to expose a whole population to death (1978: 136). Thus, while the concept of 'humanity' is a philosophical abstraction, it is one that refers to the real political situation of entire populations in the geo-political machinations of transnational power shifts. In order to understand the rise of bioethics in this context it is illustrative to consider a definitive moment in its history: the acceptance of the idea of brain death, in addition to the conventional criteria of cessation of heartbeat and breathing (Stevens 2000: 98).

The limit of death

In 1968, after years of controversy, the Hastings Center in New York City sanctioned a new set of criteria for death that had been elaborated the same year by the Ad Hoc Committee of the Harvard Medical School to Examine the Definition of Death (Stevens 2000: 81). 'Brain death' is defined as a state where the brain is massively damaged or 'dead' but the body continues to live by virtue of artificial means, for example by means of a ventilator machine (Stevens 2000: vii).[33] Brain death is when the anatomy of the body is 'in motion' but not purposively directed towards living in any way (Agamben 1998: 186; Dean 2004: 172). The new definition of death did not replace the old standard; from 1968 on there have been two forms of death under the law (Stevens 2000: 81).

The few histories that have been written on bioethics present a widely accepted view of its history, which is that bioethics, including the brain death criteria, is the natural product of the novel questions about life and death that followed after developments in medical technologies (see Rothman 1991; Jonsen 1998). However, in a more recent critical re-examination, the focus is widened from the individuals involved in the debates to an analysis of the connections between developments in technology, bioethical responses to those technologies, and the ambiguous political and professional positions of bioethicists (Stevens 2000). In *Bioethics and America*, Stevens argues that the real impetus for the definition was twofold: the need to mitigate conflicts of interest for doctors 'when deciding whether to "pull the plug" on potential organ donors who were irreversibly comatose', and the need to calm public fears about the newfound ability of medical science to sustain the 'living dead' (Stevens 2000: xiii).

There were significant legal implications to the conflict of interest faced by doctors between the desire to move forward with experimentation in the field of organ transplantation and the duty to patient health and well-being. The best source of transplantable organs was severely brain-damaged individuals on respirators who, according to the conventional criteria for death – 'the cessation of heartbeat and breathing' – were technically still alive. A redefinition of death was necessary to avoid the criminalisation of efforts to procure organs. As Stevens elaborates:

> The immediate cause that motivated physicians to establish brain death criteria and to attempt to codify them into becoming medical custom was that such a move provided something that the courts and legislatures could eventually validate as 'legitimate' under the law.
>
> (2000: 107)

As the editors of the *Journal of the American Medical Association* wrote, unless the medical definition of death was changed, a physician who removed an individual's still-beating heart might be seen as a murderer from the perspective of law, and the transplanter as a 'surgical criminal' (*Journal of the American Medical Association* 1968: 220). Thus, a new form of death was born – the 'heart-beating cadaver' – an existence straddling the conventional border between life and death. In this sense, brain death was more a legal necessity than a scientific one (Starzl 1992). In other words, the brain death criteria gave protection to physicians and assisted in advancing biomedical research.

The matter would be put to the test in 1976 in the Karen Ann Quinlan case, heard by the New Jersey Supreme Court. In 1975, twenty-one-year-old

Karen Quinlan fell into a coma of unknown origin. Doctors diagnosed her condition as 'chronically vegetative' (Stevens 2000: 110). Despite her parents' request, doctors subsequently refused to take her off the mechanical ventilator that was assisting her breathing. Litigation resulted from the doctors' refusal to 'pull the plug' (Stevens 2000: 129). At issue was the fact that Quinlan did not meet the technical criteria of brain death because she still possessed some minimal brain function. For this reason all the doctors who examined her agreed that she was not technically 'dead', even though she was virtually dead in the eyes of her family (Stevens 2000: 113). She existed in a legal no-man's land, biologically living but lacking the quality of life: between two deaths. This resulting situation is not dissimilar to the one that Agamben's concept of bare life is meant to capture.

Quinlan's form of life was in essence created by the new definition of death or, more accurately, she crossed into a threshold space opened up by the definition of brain death; what was at stake was who would control the fate of that form of life. What was at issue in the legal case was who was going to control the fate of Quinlan's bare life. The ensuing New Jersey Supreme Court decision *In Re Quinlan* elaborated that doctors should have the discretion to remove chronically vegetative patients from ventilators without legally committing murder, even if the patients did not meet the threshold of brain death (Stevens 2000: 131). The outcome of the Quinlan case was touted as a victory for a patient's 'right to die', but Stevens points out that what was also decided was that doctors had the power to make life-and-death decisions without criminal liability. Thus, the brain death criteria had created not only a new basis for heart donor eligibility, it also created a 'new generic class of existence' (biological living without life) between itself and life at the junction of law and medicine (Hyland and Baime 1976: 37). In other words, the case also concerned who controlled the fate of this legally recognised biological existence; under appropriate circumstances, doctors may kill such a life but not murder it.

In this sense, the criterion of clinical brain death was more flexible and in tune with the demand for transplantable organs and other developments in biomedicine and biomedical technologies. For this reason, one physician at the time called the brain death criteria the 'new definition of heart donor eligibility' (Stevens 2000: 87). However, it also had another effect, which was the opening up of a space of indeterminacy between life and death, which an expert knowledge could resolve. The role of bioethics, in ushering in the notion, was important in stemming potential legal liabilities for physicians and researchers; it also helped shift cultural understandings of death from an absolute limit to a potential state of being.

As with the development of brain death itself, the rise of bioethics is typically described as a natural consequence of new biomedical advancements; however, the redefinition of death was not an inevitable outcome of the invention of mechanical respirators. Stevens argues that without the professional guidance of bioethics experts, public reaction to the new criteria would have become a problem (2000). The legitimacy of the bioethical paradigm is connected to the practical role that bioethics reasoning plays in negotiating such debates. Furthermore, bioethics established its own institutional legitimacy by helping to calm public fears about the newfound ability of doctors to sustain the 'living dead' and providing an ethical framework that allowed doctors and researchers to escape the difficult moral problems associated with transplantation and the end of life (Stevens 2000: 159).

Unlike more radical critiques such as those of Mumford and Ellul, who were 'mistrustful' of biomedicine's under-girding role in a technological society, bioethicists spoke instead of 'man's' confrontation with 'politically disembodied' ineluctable processes (Stevens 2000: 1–2). The use of terms such as 'humanity' and 'nature' obscures the political dimensions of the new form of power Jonas described. Sociologists have identified something similar with the 'biomedicalisation' of society, which is associated with innovations in molecular biology, biotechnologies, transplant medicine and other new medical technologies (Estes and Binney 1989: 588; Cohen 1993; Clarke *et al.* 2003; Dean 2004). With these developments, the field of applicability for bioethics, life itself, has expanded; it is the rationality that becomes applicable when biological existence beyond human existence acquires legal recognition. In this sense, bioethics is a form of thought that emerges internal to biomedical domains in response to the problems posed by ethical issues for professionals and administrators in the course of life-and-death calculations.

In political theory, the impetus behind the development of the criterion of brain death is also often assumed to be the taken-for-granted idea that it is a natural consequence of the invention of mechanical respiration (Norris 2005). However, what was at stake in the Quinlan case is perhaps not dissimilar to the form of sovereign power that Agamben connects to the concept of bare life – in fact, he mentions the case in one of his books on sovereignty (1998). Considered in relation to the legal theory that connects the authority of law to the common good, the rise of an ethical logic that takes the existence of a moral property common to all humans – 'life' itself – as a first principle is significant. As Jonas' philosophy suggests, bioethics is based on the legal capture of the space in between brain death and a life worth living.

Between two deaths

Following Foucault, Agamben argues that the value of life is essential to the operation of power in contemporary political orders. Agamben develops Foucault's account of juridical power in order to articulate a conception of sovereignty that is not localised in a legal order but permeates all areas of social life; biological existence is now reflected in political existence, rather than being the inaccessible substrate of political existence (Foucault 1978: 142). Agamben generated the concept of bare life to capture this development in political theory. The distinguishing feature of bare life is that it is wholly constituted by its legal status. It is a legal being before it is a human being and only to the extent that it is a living being. Law exists in tandem with bare life. Agamben writes:

> We enter the hospital room where the body of Karen Quinlan or the overcomatose person is lying, or where the neomort is waiting for his organs to be transplanted. [Quinlan's] life is maintained only by means of life-support technology and by virtue of a legal decision. And yet since life and death are now merely biopolitical concepts … Karen Quinlan's body – which wavers between life and death according to the progress of medicine and the changes in legal decisions – is a legal being as much as it is a biological being.
>
> (Agamben 1998: 186)

The power of killing, which Foucault associates with sovereignty, is, in a medical context, not dissociable from the power to foster life. The decision on the value (or non-value) of life is not only made by what would be recognised as sovereigns operating in obviously political contexts in political philosophy, but also in hospital rooms, lawyers' offices and bioethics committees (Agamben 1998: 137). These decisions are examples of political power insofar as it is the limit between life and death itself that is at stake in the course of the calculation.

If biopolitics assumes 'life regulated by ordinances', it does so via the spectre of complete annihilation of life itself (Benjamin 1968: 287). Death remains a pervasive 'ghostly presence in the life of civilised states' (Dean 2004: 18). With the redefinition of death, sovereignty also manifests itself in the power to not allow dying. Slavoj Žižek explains the dynamic in these terms:

> The 'truth' of [biopower], hidden beneath the bar, of course, is power, i.e. the constitutive lie of [biopower] is that it disavows its performative dimension, presenting what effectively amounts to a political decision based on power as a simple insight into the factual state of things.
>
> (Žižek 2004c: 394)

In this sense, bioethics is the elaboration of a 'network of knowledge' that supplements sovereignty at the limit of life where its legal categories lose meaning (Žižek 2004c: 393).[34] The significance of the image of Antigone in this context is her embodiment of this limit 'between two deaths', which brought into crisis the possibility of clearly distinguishing between what is inside and outside the political (Agamben 1998: 25).

In biopolitics the humanity of the living person is an ethical question, but also a limit that determines the application of the law. Sovereignty is the authority 'to decide who shall be removed from the law' (Norris 2005: 9).

What Lacan called the ethical dimension between life and death corresponds, therefore, to what Agamben argues is a state of exception: a position where 'pure bare life, entirely controlled by man and his technology' appears directly (1998: 164). The limit in question is not that of divine *até*, but of bare life entirely subject to the force of law. The heartbeating cadaver, bodies kept 'warm, pulsating and urinating' by feeding tubes, or someone with alien desires, show us the zone directly (Agamben 1998: 164; Susanne *et al.* 2005: 6). We no longer need a mythical tragic figure like Antigone to show us 'what happens when one doesn't project one's dreams there in a sublimated way' (Lacan 1992: 216).[35] This is why Jonas' entire philosophy is dedicated to elaborating our responsibility for the survival of human life *in toto*. The infinite existence of humanity was certain for Kant and even classified by him as an 'idea of reason' like the existence of God, but this is no longer tenable.

For Aristotle, the self-revelation of life was the foundation of the good and it still is for Jonas, albeit in a radically different way. The basis in nature that was part of the ancient definition of human life remains integral to the view of life in bioethics, but with some important differences. In the ancient world, humans were distinct from animals in their reasonable natures and desire for happiness derived from the pleasure of rational contemplation. In this sense, the human was an intermediary in the metaphysical order, whose limits were related to natural and unnatural forms of desire and pleasure. Thus, unnatural forms of the good were not part of the moral domain. This was not formulated as exclusion so much as a threshold located in nature of a realm of transcendent powers.[36] Thus, the category of the human appears to be 'situated relative to the function of nature'; however, for the two sides of the division to 'become heterogeneous' they must be 'situated in the historical domain' (Lacan 1992: 211).

Primum vivere is a Latin phrase translating as the primacy of life; it refers to life as an existential quantifier for human action. This is one way to understand the Lacanian position that the value of life is a limit in the 'service of the good' (Lacan 1992: 309). The biopolitical perspective conceives the *primum vivere* as a form of power, an injunction of having to be. Foucault's elaborations suggest that politics is therefore not a matter of the good life but of the limit between life and death, in relation to which forms of life are recognised by bioethics rationalities. Bioethical expertise is required to care for life in the context of biopolitics. Biopolitics is a term that applies when biopower comes into contact with territorial sovereignty as conventionally conceived. Through biopower law becomes generative and able to foster life in individuals and populations. At the same time, the authority of law is less localisable in a state or territory and more normative and ubiquitous in the social body. These notions add political context to Jonas' ontological claim that life clings to life, for it suggests that this imperative is a limit to action with a beyond.

In what sense is the political ontological? The answer will be developed in the next chapter. However, it begins to be discernible if we consider the following comments by Slavoj Žižek in reference to the argument for imposing limits on biogenetic manipulation in the name of 'old conservative' values, like individual autonomy (Žižek 2004a: 126). Žižek elaborates Habermas' position on the need to safeguard ineffable human qualities and then suggests that it is anachronistic. It adheres to a view of nature that no longer applies:

> Does the very fact of the possibility of biogenetic manipulations not retroactively change the self-understanding of ourselves as 'natural' beings, in the sense that we now experience our 'natural' dispositions themselves as something 'mediated', not simply as something immediately given but as something that can be in principle manipulated (and is thus simply contingent)?
>
> (2004a: 125–26)

In other words, the thing that the ethical position is trying to protect from harm does not exist in the context of the current situation. This is something experienced beyond the limit of life. The sheer objectness of life is the 'ultimate traumatic point of biopolitics, of the politics of the administration of life' (Žižek 2004b: 509). As Žižek elaborates:

> Although we know that our dispositions depend on … meaningless contingency, let us pretend and act as if this is not the case so that we can maintain our sense of dignity and autonomy … [A]utonomy can only be maintained by prohibiting access to blind natural contingency that determines us [the *ex nihilio*], that is, ultimately, by *limiting* our autonomy and freedom … Since the results of science pose a threat to our (predominant notion of) freedom and autonomy, one should curtail science … The price we pay for this solution is the fetishist split between science and ethics … This prevents us from confronting the true question: how do these new conditions compel us to transform and reinvent the very notions of freedom, autonomy, and ethical responsibility?
>
> (Žižek 2004a: 126)

Speaking in a more conventional philosophical language, Jonas suggests that forms of ethical thought that are based on limits to human knowledge and power are no longer justified. No longer can we rest on the fact that in the last instance matters of life and death are in the hands of God, or some other realm beyond our power. It is no exaggeration to say that God has come to earth.

In this sense, unlike the Habermasian position, bioethics does accept that this is the situation. In the words of one bioethicist: 'In a world where the power of the atomic bomb has been tested and new genetic technology can shape future organisms, the obligation to future generations has become a moral concept' (Galgay 2007). This echoes Jonas in that we do not have an 'existential right to ignorance' of our 'new role as creator' (Jonas cited in Schirmacher 1987: 212). Lacan made a similar point by saying that what was once only evident in mythic form is now a tangible reality for humanity as a whole; the 'second death' – the mediation of the life of the species itself, not just of particular individuals – is now a foreseeable outcome of human power:

> The possibility of a second destruction has already become a tangible reality for us, including the threat of anarchy at the level of the chromosomes of a kind that could break the ties to given forms of life. Monsters obsessed a great deal those who up to the eighteenth century still attributed a meaning to the word 'Nature'. It has been a long time since we accorded any importance to calves with six feet or children with two heads. Yet we may now perhaps see them appear in the thousands.
> (Lacan 1992: 232)

Lacan is referring to the development of nuclear technology, which was reflective of the time in history in which the seminar was delivered (1959–60). Today, one could invoke the challenges posed by climate change, and in fact 'eco-bioethicists' do just that (O'Neill 2002). One could also mention research in neurotechnology and the development of synthetic biological assemblages, such as 'animats'.[37] Thus, it may have been the case that for Kant human life was an ineffable part of nature and 'divine form' beyond knowledge, and for the Greeks, that the law's life was *diké*, which was natural and divine.

However, in distilling the first principles of the bioethical paradigm from the work of Hans Jonas, this chapter has demonstrated that the bioethical attention to the 'sanctity of life' depends on the power to control the process of dying. Insofar as bioethics assumes a biopolitical definition of life, I would argue that it emerges in relation to an imperative to manage its problems, not critically analyse their causes. Thus, if biopolitical sovereignty must be understood in terms of bare life, the promulgation of bioethics should be understood in terms of this political situation as well. In this sense, it is a form of biopolitical jurisprudence, less a morality and more a technique, apparatus or rationality that connects legal authority and action outside the rule of law.

Chapter 5

Law and life

This chapter shifts perspective from the value of life to the function and meaning of life in the biopolitical. The preceding chapter argued that this form of politics is not constituted by a rule of law so much as by power over bare life in a permeating and diverse realm of biopower. The analysis of bioethics as a form of jurisprudence revealed there to be a death-focused aspect of law in a biopolitical situation where life becomes both an object and a primary value. This chapter delves deeper into the relationship between law and life than the bioethical approach. The work of Giorgio Agamben will be useful in this regard in several ways. It develops a theory of biopolitics in response to some of the most pressing issues in legal theory currently: a so-called crisis of sovereignty and state-based law in the context of globalisation and the rise of powerful non-state actors, the geo-political realm. His thesis is also distinct in that it dissociates sovereignty from the nation state. In short, Agamben argues that the rule of law at the national level is essentially breaking down and that national states are increasingly operating in 'states of exception' where the rule of law does not obtain, but legal authority – or law's *force*, i.e. sovereignty – continues to operate in a relatively arbitrary fashion. (Here, it is useful to note how he refers to the etymology of the word 'sovereignty' in German – *Gewalt* – which is translated in English as both 'violence' and 'sovereignty'.)

Understanding the way he formulates the relationship between law and sovereignty is essential to assessing his contributions to political and legal theory, for while evidently Agamben's theory depends to quite a degree on Foucault's definition of sovereignty and biopower, in Agamben's version, sovereignty is not the historical form of power but the ontology of law as such, something which is today no longer localisable in the king, the state or the rule of law. Specifically, he argues that the relationship between sovereignty and law has an essentially biopolitical structure: it is 'based on the body exposed to death' (1998: 122). It does not involve an absolute limit but depends on 'the enrolment and the filing away of the most private and incommunicable aspect of subjectivity: I mean the body's biological life' (Agamben 2004). Received political concepts must be 'revised without reserve' in light of the proliferation of such procedures across the globe (Agamben 1998: 12).

One of the more important of these categories is that of rights. Referencing Hannah Arendt on rights being continuous with citizenship, Agamben discusses how legal rights depend on the decision between outside and inside the law; thus, the meaning of rights is tied to 'the fate of the modern nation-state in such a way that the waning of the latter necessarily implies the obsolescence of the former' (Agamben 1996: 161). He points out that even discourses of human rights depend in the last instance on national forms of sovereignty for juridical status (1996: 161–62). It is also informed by debates around the shifts in the political and scientific meanings of life and death discussed in the previous chapter. Agamben's concern is with the underlying political structure of sovereignty that is evident in the promulgation of states of exception. In this sense, his approach is rather different from other legal theories that call for limiting national sovereignty in response to the challenges of geopolitics through international law and supranational institutions (Ignatieff 2001; May 2005).[38] While we are in the midst of an 'unstoppable decline of the nation-state and the general corrosion of traditional political-juridical categories' (Agamben 1996: 16), the implication of the internationalisation of law is a 'crisis' for the nation as a sovereign political order, but not necessarily for sovereignty *per se*.

While the implications of his thesis are profound, critics have expressed confusion over apparent irreconcilabilities in his text. Indeed, one way of approaching Agamben's position is to consider the controversy it has generated. For example, in a review of the original Italian edition of *State of Exception*, Antonio Negri called Agamben's critique of sovereignty fatalistic, which he contrasts with the optimism evident elsewhere in his other writing on politics after sovereignty (2003). His analysis of sovereignty of biopolitics is engaged in a 'fated … continuous confrontation with the idea of death' (Negri 2003). Indeed, one of his most well-known contributions is to suggest that a more appropriate model for conceiving our current political situation is not the Greek *polis* but the concentration camp (Agamben 1998: 123, 175–76; 2005: 2–4). Negri then notes that in other works Agamben's theory sustains 'the idea of [political] redemption' (2003). These texts are shot through with gestures towards a utopian politics, a politics *after* sovereignty, where the force of law is inoperative and law becomes something to be cited, 'played with' and even enjoyed (Agamben 1993a, 2000). In a sense, therefore, there are 'two Agambens', one who goes 'ethically and conceptually' beyond sovereignty and another who leaves us asking 'How is it possible to structure the world that [the] ontological approach constitutes?' (Negri 2003).

This heuristic inspired by Negri's review of 'two Agambens' will be used as a point of entry to consider Agamben's work in some detail. In order to make sense of the apparent discontinuity between the two, it turns to some of his writings on first philosophy and language, which do not correspond easily to the typology of 'two Agambens' (Agamben 1999). This discussion explores the value of Agamben's approach beyond the bioethical sensibility for thinking the political implications of first principles are based on life on the one hand and violence and law on the other. These elements of his work offer unique insights for addressing the connection between law and life discussed in the previous chapter. The Lacanian ethical framework already laid out is compatible, in at least a limited sense, with Agamben's conception of law. Both imply not universal principles but a threshold where moral and legal categories lose significance and the force of law remains. The affinity is evident in the work of Slavoj Žižek, whose discussions of the connection between law and violence will be used to stage a debate with Agamben's formulation of law in order to produce a synthetic account of biopolitical sovereignty. The chapter concludes with a meditation on what Agamben's theory implies about the relationship of violence and law.

The two Agambens

The previous chapter discussed Agamben's argument that sovereignty emerges through the decision on who will be removed from the law or when the law does not apply. The 'first' Agamben referred to above provisionally refers to those of his texts that are primarily concerned with this issue. Key to the critique of sovereignty is a diagnosis of the role of violence in the relationship between life and law at its heart. In this chapter, I want to explore the other side of the analysis. Agamben argues that the sovereign power to suspend the law must be understood in conjunction with the form of life that is produced in the ensuing state of exception. The term he uses for this is 'bare life'. This form of life is stripped of all human qualities except the basic fact of being alive: a form of existence that is entirely coexistent and subjected to law (1998: 164). He generates the concept through reference to a fairly obscure figure in ancient law called the *homo sacer* (sacred man). By combining a reading of ancient Roman law with selected elements from the theories of Hannah Arendt, Michel Foucault, Walter Benjamin and Carl Schmidt, Agamben argues that the concept of bare life is essential for understanding this operation of modern sovereignty.

In Roman legal order the notion of *homo sacer* designated an existence that was part of a political structure prior to human and religious law.[39] For example, a person aspiring to kinship or ruler status forfeited their right to civil protection. In this situation it is possible to kill said person without account for the citizen's right to appeal. The most important part of this is revealed in the way that the death of the *homo sacer* could not be part of a religious sacrifice nor murder under the law; this left it open to being killed with impunity and without due process. To the extent that it existed in being banned from the law, it came under the exclusive sphere of sovereignty where 'it is permitted to kill without committing homicide and without celebrating a sacrifice' (1998: 83). Agamben's most significant claim is the link between the *homo sacer* and the contemporary political sphere of sovereignty. The sovereign is a threshold position with the power to suspend the law while the *homo sacer* exists in a zone of indistinction that lacks the law's protection.

The paradox of sovereignty lies in the fact that the authority of law is, at the same time, outside and inside the legal order. As Carl Schmidt's claim highlights, 'the sovereign stands outside the juridical order and, nevertheless, belongs to it, since it is up to him to decide if the constitution is to be suspended *in toto*' (Schmitt 1985: 13). Agamben explains with reference to Arendt's discussion of the connection between citizenship, human rights and sovereignty in *The Origins of Totalitarianism* (1976) that the essence of sovereign power is the ability to decide on legal exceptions and create 'states of exception' where the law's authority is in force but the rule of law is suspended. This is the basis for the distinction between 'insiders' and 'outsiders' that minimally constitutes the rule of law. If sovereignty truly lies in the power that the legal order grants of proclaiming a state of exception and, therefore, of suspending the order's own validity, then the essence of sovereignty lies in the capacity to decide whether a particular life is a political subject or whether to withdraw legal protection – i.e. its ability to kill and/or foster bare life. The inclusively excluded status of bare life and sovereignty, as defined by Schmitt in *Political Theology* as the ability to decide on the state of exception, is inseparable (1985: 5). Insofar as sovereignty is manifested in the decision on the exception, bare life is its necessary counterpoint, hence the inevitability of the refugee, the name for one recognised by the law as being without citizenship, without legal personhood (Agamben 2005).

Against the backdrop of the critique of sovereignty are the elements of Agamben's work that Negri characterised as going conceptually beyond the state of exception. The 'second' Agamben is concerned with redeeming bare life and imagining alternatives to the sovereign form of law. In these moments, Agamben dwells not on the excluded status of bare life but rather on imagining what it would look like after having been transformed into a life after or without sovereignty (1993a, 2000). This other 'happy life' is one over which law has no hold, meaning on the one hand that law is not enforceable for its own sake but available for political subjects to use like found objects or citable texts; and on the other, that political subjectivity is not constituted by dividing humanity itself in an inclusive exclusion of bare life. Rather, political subjectivity puts its life at stake in every moment in the course of being.

One of the key texts in which these notions are elaborated is *The Coming Community*, which consists of a series of aphoristic meditations on a form of life called 'whatever' being. It is a form of life that is not related to any particular political community or subject position but rather exists as a singularity among others, all of whom are indifferent to the forms of difference that exist between them. To describe what this sphere of politics would look like, Agamben turns to Thomas of Aquinas' writings on limbo. It is worth recalling that Aquinas is one of the primary references for classical natural law theory. He presupposed the existence of God as the giver of a divine 'eternal law' behind the natural law that human beings are able to recognise by virtue of our reasonable natures. The eternal law, however, was beyond human experience. Interestingly, Agamben reconfigures the inherent deontological aspect of Aquinas' notion of natural law not as a limit to humanity but as a limit to the eternal law of God.

For Agamben, the most significant aspect of Aquinas' writing is his description of limbo, especially the way that God's judgement seems to have no force therein. The lack of a determinant relation between life and law makes limbo a state of 'privation' from the perspective of salvation because of the originary lack of the vision of God. However, Agamben highlights the ways that, despite this fact, for Aquinas, beings in limbo 'do not feel pain from this lack' (Agamben 1993a: 5). In this sense, they are 'irredeemably lost', like 'letters without destination', yet persist 'without pain in divine abandon' (1993a: 5–6). In limbo, God's law is meaningful, indeed constitutive, but it does not have the force of judgement. Relative to God's judgement, beings in limbo remain perpetually undetermined. They are 'neither blessed like the saved, nor helpless like the damned' (1993a: 6). They do not suffer because they 'do not know that they are deprived of the supreme good' (1993a: 6). Significantly, God has not forsaken these beings; rather, beings in limbo have always already forgotten God. Agamben suggests that this might be 'the most radical objection that has ever been levied against the idea of redemption' (1993a: 6). This resonates with his discussion of animal life as having a similar relation to divine judgement: they are 'saved precisely in their being unsavable' (2004: 92).

What makes Aquinas' articulations on the subject so interesting for Agamben's purposes is the way that limbo designates an internal limit zone to God's law, much like the state of exception; however, limbo figures as the inversion of the state of exception. In limbo, the force of law is inoperative even if its juridical categories remain meaningful. The reason for this is that in limbo, being itself shows 'neutrality with respect to salvation' (1993a: 6). Agamben's goal in *The Coming Community* is to articulate a version of such neutrality that is directed towards the imperative of being characteristic of sovereignty. He thus coins the concept of 'whatever being' to capture the possibility of neutrality in non-theological terms, going back to Aristotle to describe it as a pure 'potentiality', simultaneously abiding the possibility and impossibility of being (1993a: 37). Remaining unstructured by a natural purpose, 'whatever being' goes beyond the state of exception by being 'capable of its own impotence' (1993a: 35).

Thus, the idea of the coming community is of a situation where we might all live in a kind of political limbo, where life is not subject to sovereign decisions or identity-based political subjectivities. Without the recourse to coercive force, law could be made available for political subjects (singularities) to do with whatever they want (Agamben 2000: 64). Under such circumstances, Agamben suggests that far from being subject to sovereign decision, we could even play with the law if we wanted, as though it were a yo-yo or an object found on the ground (2000: 88). In other words, if bare life is a plaything for sovereignty in the state of exception in (the first) Agamben's critical works, in the utopian writing the tables have somehow turned. Appreciating this is essential to adequately comprehend Agamben's project; to focus exclusively on one side or the other produces confusion. For example, one critic writes: 'I find Agamben useful in understanding the form of violence that the sovereignty of law requires; it is curious that he does not explicitly link it to the problem of evil' (Caton 2006: 118). It is true that in his critical writings Agamben does not link sovereignty and 'evil' in a theologico-moral sense. However, in the utopian writing he addresses the question of evil, and it turns out that it is defined precisely as the violence that sovereignty requires.

Thus, far from not linking the violence of sovereignty to evil or absolute moral badness in some sense, Agamben suggests that the very constituting and constitutive basis of the sovereign form of law is 'evil'. This is called the 'power of being' in *The Coming Community*, and if we consider it in conjunction with the concepts of the state of exception and bare life already described, it becomes possible to see that it is directly associated with sovereignty. Insofar as sovereignty regards the potential for non-being to be a fault, a demonic element that must be repressed, 'evil' in the coming community consists in the 'decision' that excludes this non-being (1993a: 44):

> It is in every being that exists, the possibility of not-being that silently calls for our help (or, if you wish, the devil is nothing other than divine impotence or the power of not-being in God). Evil is only our inadequate reaction when faced with this demonic element, our fearful retreat from it in order to exercise it – founding ourselves in this flight – some power of being.
>
> (1993a: 32)

This corresponds to how the inclusive exclusion of bare life through which sovereignty 'seeks to transform itself wholly into life' is described in his critical writings (Agamben 1998: 187). Sovereign decision is here cast as a fearful retreat from the potential for non-being, a form of action that aims at self-actualisation by conserving, even monopolising, the potential to not be as 'a substance and a foundation beyond existence' (1993a: 44). The capture of bare life in a zone of indistinction relative to law is a way of making its own happening necessary by essentially transforming life into law in the state of exception. In this way, sovereignty removes the possibility of itself not occurring (Docherty 2006: 147). Thus, the imperative of being that conditions sovereignty is based on a foreclosure of the potential non-being of the authority of law itself, something then externalised in the bare life of individuals and populations. Sovereignty's way of being is the very basis of 'evil' in *The Coming Community*.

Some implications of this idea are explored in the section below; however, initially it seems only to return us to the original question. For the association of sovereignty with evil is parallel to the association of 'whatever being' with bare life. This is not a scale change, but rather law that can 'only grasp human life in the figure of bare or sacred life' (Agamben 1998: 133). It makes sense that Agamben's 'positive idea of redemption' invokes a notion of life defined as pure potentiality, a notion derived from Aristotle. Pure potentiality is simultaneously the potential to be and not be, a form of life that does not depend on an exclusion of non-being. 'Whatever being' is pure potentiality retains the potential to not be within itself. The 'good' is no longer, as it was for Aristotle, a *happy* life in the ancient sense, but life purposive towards its own biopolitical elaboration. For this reason, Agamben states that the sovereign power could maintain itself indefinitely in the form of the state of exception (1998: 47). Furthermore, relative to sovereignty all human life is potentially bare life: 'if today there is no longer one clear figure of the sacred man, it is perhaps because we are all virtually *homo sacrii*' (Agamben 1998: 115). There is a certain fatalism evident in this conception when Agamben argues that bare life cannot be freed from the sovereign relation because 'bare life is a product of the machine and not something that pre-exists it' (2005: 87–88).

Thus, the mechanism of change between sovereignty and the coming community is still not immediately clear, even with the connections above in mind. Another points out that there is no 'coherent narrative of how it is possible to move from the contemporary state to Agamben's ideal community' (Passavant 2006: 149). Agamben simply writes enigmatically that for the situation to be different the 'body that is bare life must … be transformed' (Agamben 1998: 188). At one level we could say that if law is based on being exposed to death, then the relation of law and life is determined by the limit of existence. In this sense, the relation between life and law that constitutes sovereignty is more than a legal abstraction; it is the ontology of being (Agamben 1998: 55). The next section of the chapter takes up the case of the missing theory of political action with this in mind. It begins by turning to Agamben's work on language and first philosophy, for he maintains that the way that human beings dwell in politics is the same way that they dwell in language. This is the key idea that will explain the homology between sovereignty and the coming community, and it also opens a window on the question of political action. The questions are: how neutrality might constitute a form of political power, how such a form of life could be brought about in practice, what it means for life to have the capacity for non-being, and in what sense this capacity would represent a challenge to sovereignty. The relationship between ontology and politics implied in his theory of biopolitics will then be explored more deeply, which will ultimately explain the reason a theory of political action is not forthcoming.

Politics and ontology

A basic proposition of Agamben's project is that it necessarily involves 'thinking ontology and politics' together (1998: 47). The way political subjectivity is constituted in sovereignty corresponds with how living being is determined in language: 'living being has logos by taking away and conserving its own voice in it, even as it dwells in the polis by letting its own bare life be excluded, as an exception, within it' (1998: 8). Like Lacan, Agamben takes language to be the unconditional category of perception and being; thus, both politics and ontology are linked to how human beings are constituted in language. In *Homo Sacer*, he writes that politics 'occupies the threshold on which the relation between living being and *logos* is realized' (1998: 8). In *Language and Death* he writes, 'man is the living being who, in language, separates and opposes himself to his own life, at the same time, maintains himself in relation to that life in an inclusive exclusion' (1991: 84). In other words, how we live in language is parallel to how political subjectivity exists in sovereignty. The 'primordial character' of being in language and in sovereignty involves a constitutive lack: a lack of sayability on the one hand and of bare life on the other (Agamben 1991: 45). Bare life is the foundation of political subjectivity, which is not accessible to the being that it inhabits.

In an essay entitled 'The Idea of Language', Agamben writes that 'what previous generations called God, Being, Spirit, Unconscious, appear to us as what they really are: names for language' (1999: 45). In language we are 'completely abandoned by God' (Agamben 1999: 45).[40] With the death of God 'there is no longer a name for language' (Agamben 1999: 44). This 'unsayable' foundation is a lack that is constitutive of being, a notion in compatibility – at least in a limited sense – with Lacanian theory. Indeed, Agamben claims that this view of language is a familiar point of agreement among many philosophers and political theorists and that the 'irreducible priority of the signifying function' is the foundation of being (Agamben 1999: 43–44). This relation to language is seen as a first principle. Agamben's idea of the relationship between language and being is drawn in part from Walter Benjamin's essay 'On Language as Such and the Language of Man', where Benjamin suggests that only a divine language could express the *way* that speaking being (human beings) exists in language (1978). Human existence depends on making meaning through language but we do not know the reason that this is the case. Agamben argues that the foundation of human existence is this very 'lack of foundation' (1999: 33). Thus, while language is the 'medium' of being for humanity, its 'weakness' is that it cannot bring the form of our relation to the medium to expression (Agamben 1999: 33).

Agamben connects the weakness of language that Benjamin discusses to sovereignty, whose weakness is signalled by the role of violence as a means of making the tie between life and law bind. From this perspective, the inseparability of bare life and sovereignty is the latter's weakness. Politics, he argues, is on the same threshold where the relation between living being and *logos* is realised. In other words, 'sovereignty' is also a name for how we live in language – following Benjamin we would call it our 'way of meaning' – the very form of relation 'between the human being and the linguistic' that makes political subjectivity possible (Agamben 1993b: 50). If it is true that law has renounced its breath of life and taken life as its object, which Agamben's analysis of sovereignty would seem to confirm, then what are we to make of his faith in an afterlife of the law? Despite the violent and abyssal nature of sovereignty, Agamben does also argue that it is only in the state of exception that 'it becomes possible for human beings to experience their own linguistic essence'; to bear witness to bare life is to 'encounter our own linguistic nature inverted' (2000: 115). What are we to make of this connection between language and politics for resolving the problematic that was laid out in the previous section, which related to the relationship between law and violence?

To broach this matter let us once again invoke the issue of the missing theory of political action in Agamben's work. While he may not articulate an explicit theory on the subject, he does write that the only 'truly political action' is one that severs violence from law (2005: 88). The utopian aspect of the coming community is premised on the idea of a space where the separation of law and its use of violence as a means has occurred. In other words, the world in which bare life might 'turn into an existence over which power no longer seems to have any hold' is one wherein the coercive element of law has somehow been rendered inoperable or has been eliminated (Agamben 1998: 153). This implies at least two things. First, political action is not about freeing bare life from law through legal reform, nor having it fully included in the law, under the rubric of human rights for example. Second, violence is excluded as a possibility for political action, be it against the state or with the goal of replacing an unjust law with a new one. The defining feature of the utopian elaborations is the absence of violence, resulting in a form of life that relates to law through indifference, not coercion. The implication of ethical approach to the relationship between law and violence is an impasse to political change. My perspective entails an appreciation of the utopian impulse but does not misunderstand it as a theory of political change, while taking Agamben's conceptualisation of the relationship between life and law in sovereignty seriously. For these reasons, I do not see the observation that there is a missing theory of political action to be a relevant criticism, but rather to be expected.

As mentioned above, whatever being is the fantasy of the redemption of bare life in a state of exception, achieved when it is somehow no longer kept at a distance from political subjectivity but united with it in a form of life that the force of law cannot penetrate. In the state of exception the political subject experiences the law's inscrutability as his or her own lack. In this sense, the idea that life and law are contiguous is playing two scenes at the same time, one that emphasises the violence of law that covers over the weakness of sovereignty (its inability to abide the potential for non-being) and the other emphasising the pleasure of law without lack. The gap between the two Agambens now appears to have something to do with the violence integral to sovereignty.

Violence and law

The mystery of Agamben's pronouncements on a coming politics lies in appreciating that they are utopian: it is a world where the evil of sovereignty is gone. Thus, Agamben presents not theories of political action but allegories of a world where the intimate relation between life and law characteristic of the state of exception is freed from the force of law. In this sense, critics are correct in their observation that the hopeful vision does not escape the form of law Agamben critiques. In that formulation, violence and the state of exception preclude the emergence of political subjectivity. The essential connection between life and law that Agamben locates at the heart of both sovereignty and the coming politics allows for a perpetual oscillation between a totalising and fatalistic critique of sovereignty and a fantasy of a coming politics where that relation would be neutralised and rendered inoperative.

On this point, I wish to introduce some comments by Žižek, which will eventually be linked to the broader issue of the essential connection of law to violence. He characterises what could be called the speaker's benefit of Agamben's position; it allows one to refuse the thought of violence in the context of political action, i.e. the utopian discourse is able to play two scenes at the same time: 'Agamben's focus on the vicious circle of the link between legal power (the rule of Law) and violence is sustained by the utopian Messianic hope that it is possible radically to break this circle and step out of it' (Žižek 2006: 264).

Essentially, Agamben's position gets to have its cake and eat it too. The totalising critique of sovereignty and utopian imagining of law without violence leads to an impasse that 'simply waits for some magical intervention' (Žižek 2007), or as it is called in *The Coming Community*, 'the event of an outside' (1993a: 67). While whatever being is an image of life in relation to law that does not involve violence, the critique does not seem to believe it is possible to 'break the cycle' of law and violence and radically step out of the entire circle (Žižek 2006: 264).

Indeed, sovereignty is an 'evil' ground from Agamben's perspective. In the state of exception, life and law are indistinguishable and law is in force but the rule of law is without significance. Relative to this form of law, in states of exception human life lacks any attributes besides the bare fact of biological living, which becomes the law's point of application, meaning such exceptions are abandoned by the rule of law but left totally vulnerable to legal violence that is not sanctionable. Therefore, the recognition of violence as the binding device between life and law in sovereignty, in conjunction with a refusal of sovereign action as 'evil', accounts for the fact that there is no practical theory of political action in his work. Moreover, it is the key to understanding how the diagnosis of the state of exception can give over to the thought of the coming community and politics after sovereignty. The latter is the result of the absence of the violence that defines the former. The contradiction in his work comes from the fact that while he suggests the possibility of a 'coming politics' after sovereignty, the political situation he describes seems to leave no way out of the violence unleashed by the force of law. The rejection of violence is what gives his writing on politics after sovereignty a utopian dimension. The 'coming community' is an allegory of the state of exception without the violence; that is, the force of law is inoperative instead of the rule of law. This crystallises that the idea of 'whatever being' is not politically transformative but a limit to the abyssal violence of the state of exception.

This chapter has described how Agamben's conception of politics is as an ontological threshold where life is realised and the force of law constituted. Relative to this limit, bare life becomes a 'mutable and virtually infinite survival' (Agamben 1999: 155). The 'heartbeating cadaver' is a figure of bare life in this sense; it is dead but retains a form of existence that opens on to a legal no-man's land between life and death. If legal authority is inextricably bound to the use of violence, the relation to bare life is itself a biopolitical imperative by which sovereignty 'realizes itself by simply taking away its own potentiality not to be' (Agamben 1998a: 46). If sovereign power operates by defining a threshold between two domains of existence, it also acts from outside these domains (Tadros 1998; Docherty 2006). I want to pause at the seeming hopelessness this implies, which is why the next chapter begins by considering more closely the connection between law and violence.

Chapter 6

Law and violence

The previous chapter pushed the analysis of sovereignty beyond the natural law of survival as a rationale for legal authority and instead considered its biopolitical dimension, evident in a state of exception. The Lacanian ethical view was helpful in de-naturalising the idea of life and in re-imagining it as a limit between zones of activity in the relation of law and desire, a perspective on the good that is as new in ethics as Jonas' definition of life and the moral law. The focus enabled my examination to go beyond the 'missing' element in Agamben's conceptualisation of biopolitics that has caused a critical impasse. What was revealed was a utopian relation to violence which had to do with the role of violence in the configuration of sovereignty. This chapter will continue this line of inquiry and look at the role of violence in the biopolitical thesis, which is premised on the political nature of imperatives of survival and living well. It will consider several emerging areas of legal and political thought that are taking up biopolitical notions in different ways. Agamben's own thesis is heavily informed not only by Foucault but also by Walter Benjamin, and in particular by a short text that Benjamin wrote against the backdrop of the extreme political turbulence of post-World War I Germany, on the relationship between law and violence.

Benjamin's text 'Critique of Violence' is a meditation on the relationship between law and violence. It begins by assessing the seeming paralysis of the dominant trends in legal thought on the question of the coercive element of law. The positivist view would regard the issue in terms of the criteria for the legitimate use of legal violence internal to a given legal order. The coercion is legitimate if it meets the criteria of legality from the perspective of a judge, for example. This emphasises the normativity of law at the expense of the brutality of law (Davies 1996; Lacey 2002: 118; see also Raz 1979). Violence is evaluated as a means to an end. If it is a legal end the coercion is justified. Benjamin points out how this opposing view of natural law also thinks of the relationship between violence and law in terms of means and ends. Thus, both approaches refer to the status of the criteria for cases of the use of violence. Benjamin felt that the positive view of law went beyond the natural law view, neither could justify violence as a means in principle regardless of the end, or whether violence, in principle, could be a just means to an unjust end (Benjamin 1978: 278).

Thus, Benjamin undertakes his critique by looking at the question of whether it is possible to discriminate the use of violence within the sphere of means themselves and without regard for the ends they serve. The question of violence as means alone implicates both its use outside the rule of law and within the bounds of legality. Law, however, always depends on violence. The establishment of law is signalled by the cessation of law-making violence, for example in the case of an occupation or civil war. The violence remains operative subsequently in the legal order in law-preserving forms, for example in the criminal law power of state. However remote law-founding violence may seem, it remains latent in the form of law itself. The means/ends relation of violence and law powers the 'mythical' machine of sovereignty and gives it a naturalised air (Benjamin 1978: 294). From a historical perspective, law is a structure or formation based on a rhythm of rising and falling violence in relation to law-making and law-preserving forms.

The essential quality of law-making violence can be seen in ancient myth. Benjamin refers to Niobe in this context. As discussed in Chapter 3, Niobe boasted to having more children than Zeus; in response her children were killed by the gods. This fate was sent not as a punishment for having violated a law but rather as a manifestation of power. *Até* is law-founding in this sense. In the modern context, law-founding is associated with an original violence that is subsequently rationalised and normalised in the rule of law and subsequently acts to suppress hostile counter-violence. The cessation of war or civil unrest is signalled by a peace ceremony and the establishment of constitutional law. Subsequent legal coercion preserves this earlier founding violence and the dialectic lasts until new forces or the forces earlier suppressed triumph and found a new law. Benjamin's position is therefore that law is inseparable from violence, but perhaps violence is not essentially connected to law. If the existence of violence outside the legal sphere can be assured, a pure immediate violence unalloyed to any ends, then there is the possibility of real 'revolutionary' action that could enact a profound political transformation (Benjamin 1978: 300).

Benjamin explores this possibility in the context of the right to strike, and more specifically the difference between a partial strike and a general strike. The former is afforded to citizens by the state while the latter is not. He begins by addressing whether it is appropriate to call a strike violent when it involves simply a withdrawal of activity. Granted, there is certainly the objection that an omission of action, a non-action, which a strike certainly is, cannot be described as violence, but Benjamin responds that the truth of this is not unconditional (1978: 291). He continues:

> And as in the view of the state, or the law, the right to strike conceded to labour is certainly not a right to exercise violence but, rather, to escape

> from a violence indirectly exercised by the employer, strikes conforming to this may undoubtedly occur from time to time and involve only a 'withdrawal' or 'estrangement' from the employer. The moment of violence, however, is necessarily introduced, in the form of extortion, into such an omission, if it takes place in the context of a conscious readiness to resume the suspended action under certain circumstances that either have nothing whatever to do with this action or only superficially modify it.
>
> (1978: 281–82)

In other words, the withdrawal of work is recognised by the state itself in the legal recognition of the right to strike as a form of self-defence. The violence of withdrawal is further evident when it is a means to an end, such as improved working conditions. The withdrawal is a means of extracting concessions after which the work will resume. In this context the non-action is a form of coercion, but the partial strike is akin to legal violence in this sense: it is an alloyed means that it is related to its end only as means. Benjamin then suggests that if the withdrawal of activity associated with a conventional strike is violence, there is a fundamental difference between that and the violence of the general or 'proletarian' strike.

Benjamin draws this idea of pure violence from the work of Georges Sorel. Sorel was a late nineteenth/early twentieth-century Marxist who rejected the idea that there was an emancipatory telos to history. He thought that faith in inevitable and evolutionary development of the mode of production was an error and instead embraced the idea of direct action and radical unionism as a means of social change. The syndicalist ideology he put forward is involved in the idea that political violence, irrespective of its end, could have a regenerative effect on society (Sorel 1999). Benjamin argued that this is epitomised in the withdrawal of activity of the general strike.

A genuine general strike has no conditions or specific demands. The violence of non-action in this context is 'pure'; it is not related to an end. This form of violence exceeds the means/ends logic of law. The withdrawal of activity enacted by a general strike is unalloyed to an end; it is means without end. As such, it suggests a wholesale transformation of the social order. The idea of non-action as a 'pure' action resonates with Agamben's concept of 'whatever being', which was discussed in the previous chapter in terms of the potential for non-being and the negation of one's actuality. The meaning of this in the context of politics was not immediately obvious, but if it is connected to Agamben's use of Benjamin's notion of divine violence it becomes even clearer. The next section will connect from the previous chapter with the concepts of a withdrawal of activity as violence. It does this through an analysis of a short story by Herman Melville, 'Bartleby the Scrivener', and Agamben's treatment of it, before returning to the question of violence. Mapping out how the figure of Bartleby relates to Agamben's notion of 'whatever being' already discussed will greatly clarify his apparent refusal of violence described in the previous chapter.

'I would prefer not to'

In Melville's story a narrator relates his experience with one of the strangest men he ever knew. Bartleby was an employee in the man's law office and one day, upon being asked to help proofread a copied document, Bartleby answers, 'I would prefer not to.' To the dismay of his employer Bartleby begins performing fewer and fewer duties around the office, always with the same explanation: 'I would prefer not to.' He does not explicitly refuse to do what he is asked; he simply responds that his preference is to not do it. This continues until he is literally doing no work at all. At that point, the narrator fires him, yet upon being asked to leave Bartleby responds with the same infuriatingly impenetrable statement: he would prefer not to. After this failed attempt to convince Bartleby, the boss remembers an associate friend who committed a murder in his private office. Although he wishes to be rid of Bartleby, the boss cannot bring himself to use force: 'turn the man out by an actual thrusting I could not' (Melville 1997: 57). Bartleby's presence has a strange power over his employer, who feels that he cannot do anything coercive to get him to leave the office, including calling the police. To escape Bartleby, the narrator eventually moves his law practice to a new location. But soon the tenants of the old offices come to him asking for help: Bartleby still will not leave. The narrator feels a responsibility to return to the office only to find that Bartleby has been taken to prison for his refusal to vacate the premises. A few days later when he goes to the prison to check on Bartleby, he discovers that Bartleby has starved to death on account that he preferred not to eat. The narrator believes he gains some insight into Bartleby when he discovers that he had previously worked in the 'dead letter' office at the post office, the department that dealt with and destroyed letters without destination (Melville 1997).

Bartleby's state of being is alluded to at the end of *Homo Sacer*. Agamben describes Bartleby as representing the antithesis of sovereignty in that he 'resists every possibility of deciding between potentiality and the potentiality not to' (1998: 47). This makes Bartleby a figure of the strongest possible objection to sovereignty. Given that the indifferent posture that Bartleby adopts extends not only to his boss but also to his own life, it makes sense that Agamben should refer to him in conjunction with the concept of 'whatever being' as well. Like Bartleby, 'whatever being' was distinguished by its capacity for non-being. He is a pure form of life in that he does not deny his potential for non-being. This is the reason for referring to Bartleby as an 'angel' in *The Coming Community*; his position turns the evil of sovereign being back on itself and thereby manifests its relation to bare life in a way that does not exclude the potential to not be but 'writes its own passivity'

(1993a: 37). However, considering the fact that Melville's story ends with Bartleby dying in prison, it is fair to ask how he can be a model for political action. For while Bartleby challenges the relation between actuality and non-being constitutive of sovereignty, the difference between the two is already rendered indistinct in sovereignty by the state of exception. Thus, the character is less a challenge to sovereignty than a testament to its seeming ineffability. Indeed, even Agamben writes that while Bartleby's form of being is at the limit of sovereignty it does not escape the relation of ban (1998: 88). In Agamben's reading, Bartleby's negation 'challenges our faith in the law's capacity to embody and administer justice' (Whyte 2009).

The violence of sovereignty is not forgotten in the coming community; rather, it is imagined in a purified form in the way that Benjamin described. It is not a means to an end but a power of existence based on the negation of actuality. Thus, at the end of *Homo Sacer* Agamben entreats readers to '*think* the existence of potentiality without any relation to Being in the form of actuality … and *think* the existence of potentiality even without any relation to being in the form of the gift of the self' (1998: 47, emphasis in the original). This interpretation of 'pure' violence from Benjamin's text is associated with Bartleby. In response, Slavoj Žižek has suggested that there is more to Bartleby's position and Benjamin's concept than the impression left by Agamben's utopian formulations:

> Agamben [presents] a 'negative' teleology, in which the entire Western tradition is approaching its own disastrous end, the only solution to which is to await some 'divine violence'. But what is Benjamin talking about? Revolution – that is, a moment when you take the 'sovereign' (this is Benjamin's word) responsibility for killing someone.
>
> (Žižek 2007)

It is possible to glean from this comment the point at which Žižek and Agamben part ways. For Žižek, Bartleby's refusal is more than a way of thinking; it is a form of life and an ontological intervention in the political. The unconditional quality of Bartleby's position lies not in its passivity but in its impassive aspect. He does not refuse this or that particular imperative from his boss. His refusal of the authority of law is more formal; he is unaffected by it and does not feel its force. He simply prefers not to live. In this sense, he does not negate something but affirms the negation of being. His refusal is his form of life. Arguing that Bartleby's withdrawal is 'a gesture which is thoroughly violent', Žižek suggests that it could be the 'underlying principle' of a political movement (2006: 381–82). This interpretation should also be understood in terms of the relationship between law and life in biopolitics, which is fundamentally entwined with biopower and life itself.

Agamben's utopian writings are indicative of a desire for law without violence, which is basically a moral position. Žižek suggests that this is a position that one might be able to occupy forever waiting for 'the objective conditions of the revolution to arrive' (Žižek 2002b: 9).[41] Thus, agreeing with the basic diagnosis of the state of exception and that biopolitical strategies are waged with life itself at stake, he writes that indeed 'in today's post-politics, the very democratic public space is a mask concealing the fact that, ultimately, we are all *homo sacer*' (Žižek 2003: 494). However, the point to be emphasised is not how the relation of law and violence commits us to hopelessness about the situation, but the exact opposite. The connection between law and violence at the heart of sovereignty does not mean that it is off limits for political action. Rather, the task today might be to reinvent 'emancipatory terror', which could very well involve a 'step outside' the democratic rule of law (Žižek 2004d: 15–17). In *The Sublime Object of Ideology* Žižek asks, 'Is liberal democracy the ultimate horizon of our political practice?' (1989: 221). Similarly, in *Iraq the Borrowed Kettle*:

> Why should the Left always and unconditionally respect the formal democratic 'rules of the game?' Why should it not – in some circumstances at least – question the legitimacy of the outcome of a formal democratic procedure? One should take the risk of radically questioning today's predominant attitude of anti-authoritarian tolerance.
>
> (2004d: 116–17)

This idea of democracy as a limit is drawn from Lacanian psychoanalysis. At this point it would be useful to recall the debate over Lacan's concept of the real discussed in the second chapter. The conclusion of that discussion was the claim that the rejection of ideals of the good is one of the most important analytical contributions of the ethics of psychoanalysis. However, it is often not recognised as having the magnitude of implications that I have suggested. In the next section I will elaborate these in more detail and contextualise the debate over violence in non-moral terms.

Lacanian political theory

A recent book by Yannis Stavrakakis is a good place to begin this endeavour, because the aim of *The Lacanian Left* is to provide an account of the role of Lacanian psychoanalysis in current debates in political theory (Stavrakakis 2007). The essential argument presented is that there is a common element among those on the Lacanian Left, which is derived from a shared reliance on Lacanian categories. Specifically, the attention to a fundamental lack understood to be constitutive of subjectivity and social life distinguishes Lacanian political theory; the concept of the real is often invoked in this context. As discussed in Chapter 3, there are many ways of describing the real depending on context; however, it is typically associated with lack, the absence and excess of which disrupts the symbolic order and spurs along identification via relations of desire organised around idealised objects. These objects stand as limits to an abyssal and potentially violent beyond that supplements symbolic structures. The strength of this approach is that Lacanian theory precludes utopianism through an 'awareness of negativity' furnished by the concept of the real, i.e. Lacanian-influenced political theory is premised on the notion of a fundamental lack constitutive of any political formation, which ideas of utopia cover over (2007: 17).[42] Stavrakakis argues that in the context of politics the real is equivalent to the limits of the ideological discourses that deny lack and sustain the appearance of legal legitimacy despite injustice (such as nationalism and consumerism). The difficulty arises, however, with the premise that Lacanian psychoanalysis is inherently progressive because of the concept of the real. He writes that because of the recognition of lack, Lacanian theory is imbued with an 'ethical attitude' of radical democracy (2007: 16), which Stavrakakis defines as 'symbolically institutionalise[d] lack' (2007: 10). Radical democracy is a form of politics that does not deny lack – as in sovereignty – but embraces the experience of the real, which can 'stimulate desire for new identifications' among people who are more progressive and just (2007: 59).

Considering the discussion of the ethics of psychoanalysis presented in the first two chapters, the characterisation of Lacanian theory as inherently progressive should appear somewhat suspect.

First, it is not just utopian ideals of the good that are repudiated by the ethics of psychoanalysis. The 'good' refers to any idea that functions as a criterion for normative judgement and delimiting point in a field of action. The ideal of 'radical democracy' as institutionalised lack can just as easily play the role. Indeed, Stavrakakis' criticisms of those on the Lacanian Left are based mostly on the degree to which they deviate from this ideal. This contradicts the primary contribution of the ethics of psychoanalysis in the context of legal and political study described here; its view of ethics as the relationship between action and desire provides a uniquely critical analytical lens but not a moral injunction. Chapter 2 illustrated this through the difference between Butler and Žižek on Lacan's notion of law and political action. Second, by equating the experience of lack with the good, its connection to the death drive and the line of sight this opens on to violence in the context of law is occluded. Thus, in Chapter 2 the issue was more than just the fact that Butler and Žižek refer to Lacan's theory yet generate opposed interpretations. What was missing from Butler's account of the real is also missing from Stavrakakis': the possibility of a radical act of creation *ex nihilio* that is implied by the very destructiveness of the death drive. The tendency to domesticate this aspect of the real is often discernible in discussions of Lacan's ethics.

For example, a recent book by Frances Restuccia (2006) argues that Žižek's interpretation of Lacan overemphasises self-annihilation and social destruction to the point that it is 'pathologically' obsessed with death and desubjectification (2006: xiii).[43] As a result Žižek simply describes 'an authentic act without paying much attention to where it leads. To Žižek in most of his work sheer commission of such an act is acting ethically, as if an ethical act has to be an ultimate act' (2006: 104). To illustrate this, reference is made to Antigone and her insistence on burying her brother. Restuccia suggests that Lacan's point was not the suicidal aspect of her action but the nature of her love for her brother (2006: 96). The recognition of her brother by the state (by funeral rites) is a 'lack' in the political community and Antigone's ethical position involves an insistence to inscribe that lack in the symbolic. As described in Chapter 2, Žižek highlights the fact that Antigone's is not only a suicide mission, something she herself acknowledges, but it is also a 'crazy' act from a normative point of view, which ultimately strikes at the foundations of legal authority.

The problem critics have with this is simply 'what sort of *value* psychoanalysis could possibly have in the practical arenas of the clinic or social change were its ethics to be defined as tantamount to jouissance and the death drive' (Restuccia 2006: 4). This is the question that also concerned Stavrakakis. Restuccia's perspective is similar to his; to have practical value the ethics of psychoanalysis must involve inscribing lack in a beneficial way to the social body, and not 'denying' it or 'covering' it over. This inscription of lack still engages with a destructive energy, but only as an occasion for restructuring desire in a more loving or just way. Under this view, for something to qualify as ethical, it must be focused on what is *good* for human subjects and for society. There must be something progressive and beneficial about ethical action (2006: 99). Lacanian ethics is defined by Restuccia as 'culture's very state of progress', which is not dissimilar to Stavrakakis' definition of Lacanian psychoanalysis as inherently progressive (2006: 150). The difficulty with these concerns is not that they should be unilaterally abandoned but rather that they are in conflict with the definition of the ethics of psychoanalysis elaborated in this book. In fact, they are exactly opposed; Lacanian ethics is defined by a repudiation of ideals of the good, even 'goods' that seem good to left-leaning intellectuals.[44]

To make this clearer, let us consider another approach to the question of moral criticism. Joan Copjec's book entitled *Imagine There's No Woman* (2002) is also concerned with the ethics of psychoanalysis and applying Lacan's theory to contemporary culture. Her study opens with a discussion of the *Antigone* that is compatible in many ways with what I have argued so far in this book. In particular, she notes that for Lacan, insofar as ethics concerns the relation between action and desire, it is a matter of the very ontology of the subject (Copjec 2002: 9). To make this argument, Copjec draws substantially on the psychoanalytic notion of sublimation. Briefly explained, sublimation was described by Freud as a psychic defence mechanism through which libidinal energy that could be potentially destructive is channelled by social forces away from negative outlets into activities that are more appropriate and beneficial. In sublimation, a libidinal investment shifts from an impossible or forbidden object to a particular object that has social value. The classic examples of sublimation are professional ambition, artistic creativity and scientific curiosity. Freud argued that sublimation was the affective mechanism powering the progressive development of human civilisation, making sublimation the only successful psychic defence mechanism (1961b).[45] In the seminar on the ethics of psychoanalysis, Lacan makes the connection between ethics and sublimation based on the dimension of social values in the equation (Lacan 1992: 207, 212).

Copjec is extending Lacan's comments when she suggests that ethics involves a certain form of sublimated desire, where the ideal object of desire is recognised to be lacking, thus leaving an empty space for a particular object. Ethics essentially concerns the gap within the subject between desire and its impossibility, and how one navigates the relation between desire and the object (Copjec 2002: 39). For Copjec, ethical action involves inscribing the gap in the form of our desire and changing it into a desire for the object 'just as it is'. Ethics is based on the immutability of desire and requires a negation of the 'egoistic self-consciousness', which involves imaginary identification with the ideal object, leaving a residue that is neutrality in relation to objects that may now simply 'stand by themselves' (2002: 39–40). The shift in affective investment introduces 'a change in the object itself' (Lacan 1992: 293). Crucially, Copjec argues that it is the annihilation of the foundation of the subject that occurs in ethical action (2002: 144).

In other words, the cause of desire is transformed from an ideal object that is impossible into an openness to radical particularity. This formulation has a Kantian overtone. Indeed, one way of understanding Copjec's interpretation is to refer to some of Kant's categories already discussed. For example, Copjec cites Kant approvingly on the emptiness of the moral law and transcendental inaccessibility of the highest good, emphasising in particular the point that 'nothing acts on or opposes the moral law from its exterior' (2002: 145). As the fourth chapter explained, the notion of the moral law involved the idea that the force of law makes its force felt by human beings from beyond the limit of human experience. Thus, we feel the pull of moral feelings but the precise nature of the good always eludes us. In this sense, Copjec is faithful to Kant on the unconditionality of the 'divine limitation'. The position of enunciation of Copjec's formulation and interpretation of Lacanian categories is premised on an adherence to this idea of a limit.

For this reason, Copjec's elaborations are useful yet potentially occlude an aspect of the ethics of psychoanalysis that I have tried to highlight in this study. That is, the manner in which objects appear in connection with larger political and legal discourses and the function of the good as a threshold. In Copjec's version, ethics is understood more as a practice of the self; such an approach cannot see the way of meaning by which objects are made available to desire.

It cannot apprehend the mode of production through which goods appear, nor intervene at the level of their production. Thus, it is not in a position to consider ethical action in relation to a broader political world in the context of law. Yet as demonstrated in the third chapter, it is entirely possible to hold an ideal of the good that defines itself in terms of particularity. The point is that the form of sublimation Copjec describes in the context of ethics takes the conditions through which goods appear as a matter of natural causation. However, as was argued in the fourth chapter, life itself is a 'prescribed good' in the biopolitical (Lacan 1992: 231).

Therefore, Copjec's infidelity to Kant's limitation connects her to Žižek's work in insisting that to deny the 'cut' between life and law has 'dire consequences' (2002b: 203). Stavrakakis was essentially expressing the same worry when he claimed that Žižek offers no way to distinguish between 'divine and satanic' acts (Stavrakakis 2005: 178). Indeed, Žižek has suggested that one of the implications of the ethics of psychoanalysis is that in order to effect political change it is necessary to risk the possibility of the 'very worst' (2000: 149–60). He argues that the ethical repudiation of the good implies that change entails a connection that appears 'impossible' in a situation – from an internal point of view at least. This is the lesson that he draws from the ethics of psychoanalysis. When the impossible happens it has the power to change the limits of what is possible in a situation in the first place (Žižek 2004b: 511). For example, during the French Revolution the impossible happened, an entire people stood up in the name of justice and equality (Žižek 1999). One critic distils the sentiments of many who have responded to this formulation: 'we should all be familiar with the criticism of the "step outside" democracy: this outside is either totalitarian or fascist' (Shaw 2005: 148). In the ethics of psychoanalysis, 'ethical' refers not to morality and the realm of social values in a normative sense, but to the relationship of action and desire. As elaborated above, clearing the moral ground puts us in a position to understand the function of the good as a threshold in an economy of desire that extends beyond it.

Can't you see that I am dead?

By referring to the death drive the ethics of psychoanalysis suspends in the realm of social values. This is not to say that the concept has nothing to contribute to political analysis. Far from it – its value lies in the very indifference to the boundaries between moral domains. In *The Ego and the Id*, Freud associated the death drive with a 'tendency to unity' in human beings that is inseparable from an 'inclination to annihilation and destruction' (1961b: 44–45). Lacan mentions *Beyond the Pleasure Principle*, wherein Freud elaborates the concept of the death drive to explain why 'there is unquestionably no universal instinct toward higher development' in humans (Freud 1961a: 41). He then explores the implications of this for ethics:

> We must see right away how crude it is to accept the idea that, in the ethical order itself, everything can be reduced to social constraint, as is so often the case in the theoretical writings of certain analysts – *as if the fashion in which that constraint develops* doesn't in itself raise a question for people who live within the realms of our experience. In the name of what is social constraint exercised? Of a collective tendency? Why in all this time hasn't such social constraint managed to focus on the most appropriate paths to the satisfaction of individuals' desires? Do I need to say any more to an audience of analysts?
>
> (Lacan 1992: 225)

Lacan suggests that in the absence of the good it is precisely the form of moral norms and their way of being that becomes important. Freud's notion of sublimation provides an idiom to connect this to ethics, but it in turn complicates Freud's concept if we see the domain of social values in relation to law. While ethics is constituted in a normative set of conditions through which objects acquire social value, it is also implicated in changing those conditions themselves. The de-naturalisation of the good in Lacanian ethics is connected to the way that the good 'has entered the political realm' (Lacan 1992: 292).

In the context of the clinic, the ambiguous way in which constraint develops is evident in an analysand's simultaneous experience of love and hate in libidinally charged relationships. The knot of law and desire so constituted is analysable with reference to fantasy structure and symptomology. However, in the context of the political, the 'fashion of the constraint' involves political processes, institutional arrangements and violence. Through law, sovereignty seeks to turn into life by inclusively excluding the potential for non-being in a social situation. The previous chapter argued that power in biopolitics involves the power to ward off death by controlling the means to foster and enhance life and the sovereign repression and exclusion of the potential for non-being. This was connected to the idea of lack at the foundation of the biopolitical that is 'identical with an order of language' (Lacan 2006: 230). Like Antigone and Bartleby, the human factor 'escapes us' (Lacan 1992: 124). Morality is thus a name not for a divine limit or natural process, but for the 'creation without a creator' (Lacan 1992: 126) and the power of 'atheistic creationism' (Lacan 1992: 261).

From a Kantian perspective, this idea may be a grave philosophical error, possibly even the worst error that leads one to diabolical evil (Copjec 1996). A more philosophical way of putting this would be to say that when it comes to the good of life, a collapse of the *is* and the *ought* is empirical. Indeed, for Agamben there was something evil about sovereign being insofar as it holds apart bare life and is not localisable in the self. The value of his analysis is the connection between law and violence that it surfaces. However, if we take Jonas' response into account, it is possible to regard the division in life itself as a natural fact, and not an error of reason or morality. From this perspective, the Kantian position in the present moment occludes the problematic that Jonas articulates, and the disturbing implication of the fact that God has come to earth and we are him (Zupancic 2000: 97).

In this sense, Antigone and Bartleby are both ways of entering the 'zone between two deaths' discussed in Chapter 2, the zone of symbolic death excluded from the social order. At this point, we can consider for a moment another possible version of living death. Lacan refers to a story by Edgar Allen Poe entitled 'The Facts in the Case of M. Valdemar', published in 1845. In the tale a narrator presents the details of the case of Valdemar. The narrator is a doctor who is interested in conducting an experiment to see the effects of hypnosis on a person at the moment of death. He is especially interested 'to what extent, or for how long a period, the encroachments of Death might be arrested by the process' (1845: 561). Ernst Valdemar is a friend who is dying of tuberculosis and it occurs to the narrator to broach the subject with him. Knowing Valdemar as he does, the doctor doubts he will have moral objections and he has no family to interfere. Indeed, Valdemar is excited by the idea of being hypnotised. It is arranged that approximately twenty-four hours before his death is predicted, Valdemar will summon the doctor.

Some months later the narrator receives a note from Valdemar and arrives at the hospital within an hour; Valdemar urgently insists that the hypnosis begin before it is too late. Upon the arrival of a medical student as a witness, the narrator proceeds with the hypnosis. Valdemar sinks into a state of hypnotic slumber as he begins to die. His extremities grow cold, his breathing all but ceases, and his pulse is imperceptible. Yet 'the general appearance was certainly not one of death' (Poe 1845: 563). The doctor is able to talk to Valdemar, who reports that he is asleep and asks to be left to die. When asked if he is in pain, Valdemar says no, and that he is dying. When the doctor then asks if he still sleeps, Valdemar says, 'Yes, still asleep – dying' (ibid.). After some minutes his body begins to contort and discolour with death. The doctor asks again if he is asleep but rigor mortis has set in more fully and the body is completely still. Then there is a tremor in Valdemar's tongue and the corpse emits hideous sounds and announces, to everyone's horror, that he is dead. The doctors are able to arrest the death process.

Valdemar persists in the state of living death for seven months. At this point the doctors decide to try to wake him. As he begins to stir suddenly 'several screams from the poor wretch are heard – "*For God's sake! – quick! – quick! – put me to sleep – or, quick! – waken me! – quick! – I SAY TO YOU THAT I AM DEAD!*"' (Lacan 1988: 231).[46] The narrator hurries to awaken him but instead Valdemar's body shrivels and within a minute is reduced to a putrefied liquid. Like Antigone, Valdemar is beyond life and death.[47] The story has some obvious resonance with the earlier discussion of bare life, heartbeating cadavers, bioethics, etc. However, it is essential to appreciate how Valdemar is an 'echo' of Antigone, in the sense of being 'between two deaths' and excluded from the social space (Lacan 1988: 231). Such a form of life makes 'actual the conjunction of death in life'; it is 'a life which is dead' (Lacan 1988: 232). In biopolitics, the naturalness of life also has an ideological function, because it functions as a limit to thinking in this domain.

The concept of the death drive is useful in this context for the way that it denaturalises the imperative of survival and the invocation of life as the common good. It describes instead the dialectic of desire for something other than continued social existence, even for something indifferent or opposed to that existence. It is precisely a gesture that risks or even pursues death. From a Lacanian perspective, ethical action necessarily involves the suspension of reasonable or 'natural' purpose. Its power does not depend on the pre-existing symbolic norms. In this context, the divine limitation in Kantian ethics remains useful but its meaning is shifted in the biopolitical context. Copjec's alternative notion of the ethics of desire obscures this beyond as well, for it is not only at the level of the subject but also at the level of law. The case of Valdemar illustrated this, as the character represented another way of entering the beyond of life and death, one which has more resonance for us because of how it prefigures bare life, the object of bioethics. Hence, two of the questions raised by Lacanian ethics are the function of life as a limit and the form of political power underlying bioethics. However, the implications include a rejection of the bioethical value of life and the Agambenian refusal of violence, which make it potentially troubling to some. It is to these issues that the conclusion turns.

Conclusion: A future uncertain

The background of this study is the profound shifts currently under way in politics, especially in the field of legal authority. The international system of law based on sovereignty and mutual non-interference that was established in Europe by the Peace of Westphalia in 1648, and which spread over the globe in various ways for the next 300 years, is being profoundly transformed in ways that are still obscure (James 1999).[48] A shift away from sovereignty is detected by an array of domains, such as international law, pressures for sovereignty from within national borders, and new types of conflicts around citizenship and immigration (Milde 2002; Lippert 2005; Besson 2006). Technological developments permit the rapid global movement of goods and persons in ways that often defy national juridical control, and at times create tension with civic nationalisms and notions of belonging associated with the nation state (Krishna 1999; Nef 2002; Painter 2002; Pavkovic and Radan 2003). The influence of international capital, the establishment of either a new international American Empire or a new era of international law, and the resurgence of religion as a dominant source of political identification among people have challenged images of democratic politics based on the rule of law as a founding element (Lindseth 1999; Keating 2001; Oliver 2003).

Serious legal circumscription to the power of nation states has also appeared in the twentieth century. Within Europe, the European Convention for the Protection of Human Rights and Fundamental Freedoms, as well as the EU, has created new layers of governance which have divided the political authority of their member states. Within the wider international community, rules governing war, weapons, terrorism, human rights and the environment have transformed and delimited the power of states with new forms of accountability (United Nations Office of the High Commissioner for Human Rights 2006). Some of these arrangements are consensual (for example, a state agrees to abide by rules that limit its policy options; trade agreements and international treaties). Others are non-consensual (humanitarian justification for the imposition of diplomatic, economic or military measures on some states by others). Nation states are less and less regarded as discrete political worlds (Corrigan and Sayer 1985). In the post-Nuremberg era of international law, the boundaries between states are decreasing in legal, economic and moral significance (Tierney 2006). This marks a significant step away from the regime of international law that existed prior to the twentieth century, which was largely defined by sovereignty. As such, it cuts to the heart of what has defined law and legal authority for centuries. The assumption that sovereignty is an unlimited, indivisible and perpetual form of power is increasingly regarded as anachronistic, and a number of commentators have argued that law itself is in the midst of a crisis of sovereignty (Turner 2003: 50; Miller 2007: vii).

As a result, the idea that a connection between law and state power is necessary or even valid in an international context is increasingly being called into question. One important contribution in this regard argues that sovereignty does not have to be abandoned, so much as rethought and redefined to fit in line with emerging international norms, principally respect for human rights. For example, Michael Ignatieff has argued that sovereignty should be 'divided' or 'shared' between nation states and international law in order to ensure that the exercise of state sovereignty is conditional to some degree on proper human rights behaviour (Ignatieff 2001). Thus, in a conflict between state action and human rights, international law trumps sovereignty in the name of the good international community. Within this framework, states may forfeit claims to sovereignty if they violate the standards and values embedded in the international order; and such violations would not be a matter of morality alone. Rather, they would be a breach of a legal code that may call forth the means to prosecute and redress it.[49]

However, the authority of international law is ultimately based on the effective power of states for enforcement. Thus, the question is where the effective power of international law, including human rights law, comes from if not from the willingness of sovereign states to enforce it. How does moral force translate into legitimate legal authority? Furthermore, the idea that human rights should function as a limit on the policy of nation states seems to potentially invert the classic law and sovereignty model, in essence redefining sovereignty itself as the duty to protect human rights. We could ask how an international rule of law would be different from sovereignty writ large. Where would ultimate authority lie? In whose name would it operate? The human rights formulation of law forces a revisiting of some of the most fundamental questions in the study of law and society, such as where law comes from and how it is justified.

Alongside the discussion of the decline of sovereignty in natural law, a supposedly Aristotelian idea of natural purpose has emerged. Confirming this as a universal morality is the nature of law and the decisive reason for the duty to obey, in national, international or indeed any political setting. Such a move reinforces the value of Aristotle and the notion of the common good, especially the idea of having a natural impulse to work together politically to attain the common good. The reason I summarised the role of happiness and the lawgiver in Aristotle is because new natural law theory posits the value of human life as a natural purpose that need not make reference to the divine or anything transcendental for corroboration. The point made in the second chapter is that natural law discourses in which nature and the divine are explicitly disassociated have a fundamentally different meaning from the ancient one. The ancient account is not as human-centric as it initially seems.

Aristotle's conception of political community presupposed the realm of the gods in a double sense. The morality it is based on comes first from a divine reference, since it was the pleasure of the gods that designated happiness and the highest good, and, second, through the superhuman virtue of the lawgiver in enacting the rule of law. In the ancient definition, nature was continuous with a divine realm on these levels. It is both a threshold and beyond to law and ethics. The difficulty for natural law theory in the contemporary context relates to the role of this ineffable externality, which supplements the notion of a common morality that undergirds political community. In Aristotle, this certainty is located in the virtuous nature of the lawgiver and the idea of the highest good, which are both, in a sense, levels upon which the divine intervenes through nature to play the role of the sovereign. In secular natural law theories, the purpose of law is seen as determinant and knowable. This definition is premised on conception of nature defined by scientific knowledge. Along with this, then, comes the belief that legal principles can be discovered by human beings – a significant contrast with classic and ancient natural law theory alike, where the source and scope of both individual morality and legal authority was understood to be beyond human imagination and direct experience.

In a sense, therefore, in the new natural law the theological aspect is not missing. Rather, it is substituted with the idea of rational nature and displaced by the certainty of moral knowledge which asserts that survival is the minimum purpose of life, which is seen as continuous with nature. From this point of view, the authority of law is purposive in relation to life, which is a universal good determined by nature. Thus, the difficulty for natural law theory in the contemporary context relates to the status of life as a moral fact and politically determined concept. For the philosophical language of natural law, like the language of bioethics, depoliticises legal authority. What must be developed in the study of law and society are methods to analyse the position of enunciation of so-called 'moral' attempts to nurture life, including discourses of human rights, without objectifying the naturalistic intuition that these discourses assume.

I used Lacan's ethics of psychoanalysis to introduce a difficulty into this tendency in the bioethics. Lacan's approach to ethics is related, on the one hand, to the relation between action and desire and the good as a limit in this relation, and on the other, to a repudiation of conceptions of morality defined in terms of rational and purposive desire and happiness. I analysed the paradigm of bioethics and argued that therein 'life' is not seen as an ineffable limit, but as an object imbued with an automatic and determinable purpose. The new reign of ethics is not, as is often assumed, the natural consequence of the development of biomedicine (Rancière 2004). Rather, bioethical rationales control contexts where the definition of life is in doubt and command authority on the basis of a claim to empirical knowledge about morality and the causal origin of moral principles like autonomy, non-malfeasance and justice. I drew on theories of biopolitics to suggest that this claim is actually a political act, a sovereign one. It was concluded from this that bioethics is less properly understood as an ethics and more as an instrumental rationality internal to the biopolitical. In other words, it is a form of jurisprudence that corresponds to a political situation where life and death are privileged and shifting legal categories. From this perspective, sovereignty is not so much in crisis as it is transforming its shape with respect to locus and way of operation. Less mediated by territory or social contract, it is exercised directly on the level of life using the imperative to live.

Therefore, to describe the current condition of sovereignty only in terms of a 'crisis' of the nation state is to understand sovereignty in a limited way – that is, only in terms of historically precedent forms of law. In this precise sense, it could be argued that Agamben takes a natural law approach when he describes the biopolitical essence of sovereignty and implicitly rejects the idea that it can be divided or shared. Thus, if a state is subject to international law, sovereignty is not a property of the state, but this does not mean that sovereignty, in some non-national configuration, is not operative. In fact, this is one of the important implications of Agamben's thesis: that sovereignty is disarticulated from the nation-state framework and no longer localisable in the state in the same way, but this does not necessarily mean that sovereignty's undivided and unlimited qualities are vanishing. Quite the opposite. they seem to be quite evident in the proliferation of conflicts around the globe in which life itself is at stake in political decisions.

Thus, the question certainly remains about the determination and enforcement of 'proper' behaviour, but more importantly so does the problem of understanding how the determination of life is also a political intervention. I referred to theories of biopolitics to argue that the form of sovereignty emerging is not necessarily connected to politics at the national level in a straightforward way. Biopolitical sovereignty is less invested in the efficacy of the rule of law in a specified territory than in the capacity to control and manage the life of populations in a territory. I developed the analysis of life as a political concept using a non-moral ethical perspective developed from the ethics of psychoanalysis. The difficulty this presents is not simply that there are particular interests (medical, scientific, corporate, etc.) that operate behind so-called universal definitions of 'life'. The problem lies in the way that bioethical discourses of life depoliticise biopower and naturalise the biopolitical.

Currently, the tendency in legal studies is to oppose sovereignty and international law. Theories of the biopolitical make it possible to conceptualise the operation of sovereignty in a way that exceeds this division. Biopolitical sovereignty is a force not necessarily wedded to national interests or jurisdictions. Attachment to the idea of the 'nation' in political science, which is evident in conceptions of limited sovereignty, ignores the fact that a limitation fundamentally transforms the meaning of 'sovereignty' in that context where properly speaking the term no longer applies. In light of the development of bioethics and shifts towards legal paradigms centred on life, the meaning of juridical categories like right and citizenship, which have traditionally been defined in relation to sovereignty, must also be revised. The Lacanian view of ethics proved to be helpful in this regard by de-naturalising the concept of the good and re-imagining ethics in relation to a dialectic of law and desire – a relation that is increasingly indistinguishable from politics.

The changes currently under way to systems of power and authority in the world involve a redefinition of national sovereignty and the emergence of new forms of global association. The proliferation of 'biopolitical sovereignty' centred on the division between life and death is probably only one possible direction developments could take. This study was intended to offer the beginning of a vocabulary for assessing new forms of power, as well as the role of sovereignty in international law. In view of the ascendancy of bioethics, the non-moral view of life is essential for critical analyses of legal authority. This fact raises several issues, including the ethical status of violence, the political determination of life, the changing nature of the nation state and the implications of democracy as a limit to political imagination. Lacanian ethics is especially useful in approaching these questions because of its repudiation of ideals of the good and non-moral focus on the relationship between action and desire. It does not delimit an analytic field based on a moral rejection of desire that is irrational, non-purposive or otherwise normatively unappealing. The ethical distance from moral claims and endlessly circulating bioethical imperatives to live well offers an alternative vision of force which 'in former days' was 'expressed in terms of the gods' (Lacan 1988: 227). This perspective sees the desire for survival in relation to a form of death 'that is precisely there under life' (Lacan 1988: 232). This point is new in ethics and normative political theories. It is a 'revolution [that] brings man back into the world as creator' (Lacan 1988: 231).

Notes

1 Fuller elaborated eight principles of the inner morality of law, which I paraphrase: laws must be expressed in general terms; they must be public; their effect must be mostly predictable; they must be generally understandable; they must not be contradictory; they must not require things beyond the actual capacities of subjects; they must be stable and not changed too frequently; finally, laws must be administered in a manner consistent with their wording (Fuller 1969: 39).

2 The Rule of Recognition lays out the 'feature or features possession of which by a suggested rule is taken as a conclusive affirmative indication that it is a rule of the group to be supported by the social pressure it exerts' (Hart 1994: 92).

3 For a discussion of other approaches to legal study, such as sociological jurisprudence and critical theories of law, see Dragan Milovanovic's *Introduction to the Sociology of Law* (2003).

4 Theories of biopolitics, including Michel Foucault's discussions of sovereignty, do not have a significant presence in the debates in the philosophy of law. A parenthetical in an entry on legal positivism in the *Stanford Encyclopedia of Philosophy* written by a prominent commentator on the Hart/Fuller debate is indicative of this. Green is concluding a discussion of early positivism when he writes:

> What survives of their [Austin's and Bentham's] outlook is the idea that legal theory must ultimately be rooted in some account of the political system … their particular conception of a society under a sovereign commander, however, is friendless (except among Foucauldians, who strangely take this relic as the ideal-type of what they call 'juridical' power).
>
> (Green 2003)

By juridical power Green means the authority of law; however, his characterisation of its place in Foucault's thought is inaccurate. Foucault argued that the 'juridical' conception of power is a vestige of monarchy. In the introduction to *The History of Sexuality* he writes that we must break free of the image of power thought strictly as 'power-law, of sovereignty, which was traced out by the theoreticians of right and the monarchic institution' (1978: 90). The implication that there is a historical transition from sovereignty to biopower has been the subject of debate among those who study Foucault and draw on theories of biopolitics. It is a 'historical form that is characteristic of our societies … characteristic yet transitory' (1978: 89).

5 In *The History of Sexuality* Foucault writes: 'The sodomite had been a temporary aberration; the *homosexual* was now a *species*' (Foucault 1978: 43).

6 The terms 'biopower' and 'biopolitics' are often equivocated by commentators (for example see Ojakangas 2005).

7 Unlike Foucault, Agamben did not feel that biopolitics was appropriately understood as either only a distinctive feature of modernity or the institutional arrangements constitutive of modern society. Rather, it is at the origins of the Western conception of the political. He turns to philosophical discourses of Aristotle and others to make the argument, while Foucault was sceptical of philosophy's ability to conceptualise historical problems (Ross 2007).

8 *The Washington Post* reports that 'military-style' roadblocks were set up around a crime-ravaged neighbourhood in Washington DC to 'keep out criminals' and anyone without a 'legitimate purpose'. The mayor explained that 'in certain areas we need to go beyond the normal methods of policing' (Klein 2008: A1).

9 Eteolces and Polyneices were to alternate the throne every year. Eteolces had refused to step down after his year was up, which precipitated Polyneices' attack.

10 This interpretation has been very influential. In Western legal theory, the play is typically read this way and the equation of the rule of law with justice often endorsed (Milovanovic 2003: 1–5). It is generally accepted that 'justice' refers to 'the legal resolution of social relations ... through general rules applied in an impartial fashion to all persons alike and known in advance ... according to the internal logic and validity of the system and excluding any personal or non-legal reasons' (Veitch *et al.* 2007: 21). In this scheme, law is seen as the embodiment of community spirit and its most substantive articulation. For natural law proponents, this is the 'internal logic' of law, which is why it is also an ethical good. For example, Fuller writes of the 'inner morality' of law (Fuller 1969: 33–44 and 145–62). On the other hand, positivists see law as a means of resolving conflicts, and the morality of law as a matter of the legality and the correct application of the legal principles enshrined in a given legal order (Milovanovic 2003: 2). Both see the role of law in society as a binding agent of order. In positive law the principle of 'legality' bestows a quasi-moral legitimacy on any action that is legal according to the legal order. The judgement of whether the action is *just* is suspended for the sake of the analysis. While natural law is also enunciated from a position that is committed to particular legal norms and willing to make the claim that an action is unjust if it deviates from these norms, even if it is legal from a perspective internal to its own legal order.

11 As Kellogg elaborates:

> Derrida's (1986) reading of a pivotal section of Hegel's *Phenomenology of Spirit* demonstrates Hegel replaces a necessary logical step with the figure of a mourning woman, the Greek mythological character Antigone, and in so doing manages to absorb into the philosophical system precisely what must always remain firmly beyond its reach: the singularity of the singular. My analysis goes to show the precise way that Hegel's choice of a singular figure – Antigone – occludes the impossibility of the encounter between the singular and the universal. Specifically, Hegel wiggles out of a tight spot in his analysis by appealing to the political prejudice that 'woman' is both representable and generalisable, and unrepresentable and singular; the divine and unrepresentable 'other'.
>
> (Kellogg 2003: 363)

12 The pleasure principle is one of two principles of mental functioning posited by Freud in his metapsychological writings. The pleasure principle involves maintaining a balance of excitation as constantly as possible by avoiding unpleasures and obtaining pleasure in measured doses. Eventually, Lacan explains, the subject 'is forced to form a conception of the real circumstances in the external world'

(1992: 252). The reality principle is a way of safeguarding the experience of satisfaction produced by the pleasure principle, achieved by taking into account the limits of the social world. The pleasure principle works with the reality principle through psychic defence mechanisms – such as sublimation – to achieve satisfaction by permissible and acceptable means (i.e. indirect or delayed gratifications). Lacan rejects the notion of reality as an objective and self-evident given, however. He emphasises Freud's point that the reality principle is still ultimately in service of the pleasure principle, not reality: 'the reality principle is delayed action of the pleasure principle' (Lacan 1988: 60). Lacan says: 'In truth, we make reality of pleasure' (1992: 225), and thus challenges the idea that the subject has a reliable means of distinguishing reality from fantasy.

13 It has been suggested that Freud and Lacan reference ancient Greek tragedy in an ahistorical way. For example, Jean-Pierre Vernant claims the 'historical moment' of Greek tragedy was in ancient Athens and nowhere else; see Vernant and Vidal-Naquet (1988). Derrida justifies referring to the classical structure of myth to shed light on 'urgent contemporary matters'. He writes: '[Myths] interest us and we take a look at them at the points where they seem, as though of themselves, to deconstruct these inheritances or the prevailing interpretations of these inheritances' (2000: 139).

14 After 'the reduction of the resistances' in analysis what remains is the 'tendency to repeat' symptomatic forms of action. This is another form of 'resistance' (Lacan 1988: 321).

15 More of the implications of this form of desire will be discussed in Chapter 3.

16 Žižek changes the way he conceptualises the real over the course of his work. In *The Sublime Object of Ideology* (1989), he describes the real as a hard kernel and an extra-discursive support of the symbolic but later the real becomes a name for the limits, gaps and inconsistencies internal to the symbolic and not an extra-discursive object. See the introduction to *For They Know Not What They Do* (Žižek 2002a).

17

> The step taken by Freud at the level of the pleasure principle is to show us that there is no Sovereign Good – that the Sovereign Good, which is *das Ding* [the Thing], which is the mother, is also the object of incest, is a forbidden good, and that there is no other good.
>
> (Lacan 1992: 70)

> The Freudian world assumes that it is the object, the Thing, as the absolute Other of the subject, that one is supposed to find again. It is to be found at most as something missed. One does not find it, but only its pleasurable associations.
>
> (Lacan 1992: 52–53)

18 While Aquinas connected God and natural law, he did so in a qualified manner:

> The plan, by which God, as ruler of the universe, governs all things, is a law in the true sense. And since it is not a plan conceived in time we call it the eternal law … Reasoning creatures follow God's plan in a more profound way, themselves sharing the planning … This distinctive sharing in the eternal law we call the natural law, the law we have in us by nature. For the light of natural reason by which we tell good from evil (the law that is in us by nature) is itself an imprint of God's light in us.
>
> (Aquinas 1989: 281)

There is much more nuance to the tradition of natural law than I can do justice to here. See Draper (1990) and Johnston (1998).

19 Altizer writes:

> Above all it is in greatest tragedies of Sophocles that we may observe an ultimate conflict between the human and the divine, and here we may also encounter the fullest embodiment of individual character and identity. It is also the fullest realisation of individual consciousness which occasions and calls forth a tragic destiny. Hence it is individual virtue or *arete* of the Sophoclean hero which necessitates the wrathful vengeance of *ate*.
>
> (Altizer 1985: 37)

20 Another interpretation suggests Antigone's action is that of a hysteric. In Lacanian psychoanalysis a hysteric subject is the agent of a discourse, which is connected to a law to which they cannot see their own subjection. This reading of Antigone argues that when she puts concern for her own life aside and violates Creon's law, she does so in a way that is 'entirely consistent with, and binds her to, her family destiny and paternal law' (Grigg 2001: 121). The significant moment is when Antigone realises what she unwittingly 'has been for the Other' – by 'Other' is meant her father and the Labdacus curse (Grigg 2001: 120). Technically the question of what one is for the Other in Lacanian theory is not directly the 'law of the father'; rather, it is a matter of the desire of the Other. As Lacan writes in *The Ethics of Psychoanalysis*, Antigone pushes to the limit something that might be called the pure desire for death. He continues:

> Think about it. What happens to her desire? Shouldn't it be the desire of the Other and be linked to the desire of the mother? The text alludes to the fact that the desire of the mother is the origin of everything [see Sophocles 1974: 149]. The desire of the mother is the founding desire of the whole structure, the one that brought the unique offspring that are Eteocles, Polynices, Antigone and Ismene.
>
> (Lacan 1992: 283)

Without going into unnecessary details about the concept of the hysterical subject in Lacanian theory, the interpretation hinges on the point that it is only *after* Antigone has violated Creon's law that she becomes aware of the meaning of her action vis-à-vis her family curse and what she has sacrificed in doing it. Her obstinacy subsequently emerges in insisting on being the bearer of this fate. The reason that we are rejecting this interpretation is that its intelligibility depends on the idea that it is only after the fact that Antigone realises what the consequences of her action will be. Our view follows Lacan's comments that emphasise the opposite: that the distinguishing feature of Antigone's situation is her prior knowledge of her fate (1992: 273).

21 Antigone was without a city by nature. In the Aristotelian universe, there were two ways of being without a city, one by chance and the other by nature. An individual without a city by chance is a human being who has the rational capacity for politics but lacks access to a city in which to develop it. For example, peasants living their entire lives in the countryside lack a city by chance and as a result never have the opportunity to develop their ethical and political capacity. On the other hand, an individual who is without a city by nature 'is incapable of community or has no need for one' (Aristotle 1962: 1253a27–29). For example, a Cyclopes is without a city by nature. It lives in a cave because it inherently lacks the inclination for politics, not because contingent circumstances deny the opportunity to develop such inclination. Antigone is not outside the city quite like a

Cyclopes, but it is possible to see her as the bearer of inhuman nature. One who cares not for her own survival or the good of others is a law unto herself (Mayhew 1997: 19).

22 This document was most recently updated in 2005. The three federal funding bodies for research in Canada, the Canadian Institutes of Health Research (CIHR), Natural Sciences and Engineering Research Council of Canada (NSERC) and the Social Sciences and Humanities Research Council of Canada (SSHRC):

> wish to promote research that is conducted according to the highest ethical standards. The Agencies have therefore adopted this Policy as [the] standard of ethical conduct for research involving human subjects. As a condition of funding, we require, as a minimum, that researchers and their institutions apply the ethical principles and the articles of this Policy.

The Tri-Council Policy Statement: Ethical Conduct for Research Involving Humans 1998 (with 2000, 2002 and 2005 amendments) can be accessed at www.pre.ethics.gc.ca/english/policystatement/policystatement.cfm

23 This criticism reflects a traditional ethical distinction in ethical philosophy between 'individualist' and 'communitarian' ethics (see Nussbaum 1986).

24 *The Tri-Council Policy Statement: Ethical Conduct for Research Involving Humans* describes the policies of the Medical Research Council (MRC), the Natural Sciences and Engineering Research Council (NSERC) and the Social Sciences and Humanities Research Council (SSHRC). The document replaces the SSHRC's *Ethics Guidelines for Research with Human Subjects*, the MRC's *Guidelines on Research Involving Humans* and the MRC's *Guidelines for Research on Somatic Cell Gene Therapy in Humans*.

25 In terms of medical and pharmaceutical research, bioethical protocols have been increasingly implemented since the Nuremberg Code, which came out of the trials of doctors involved in Nazi medical experiments (Annas 2005), and the Tuskegee experiments in the United States (Peckman 2001). See especially Chapter 5 of Annas' *American Bio-ethics: Crossing Human Rights and Health Law Boundaries* (2005).

26 More from Kagawa's article:

> The recent tendency of Balkanization in the field of bioethics could endanger the integrity of moral investigations. Subfields of bioethics, such as genethics, neuroethics, nanoethics and so on, originate consecutively, entail wastage of valuable time and money, and increase the risk of fragmentizing moral considerations in an inconsistent way.
>
> (Kagawa 2009: 7)

27 For examples of theory-led approaches that are influenced by Utilitarianism see Singer (1993) and Hare (1975). Singer often draws conclusions that some find shocking, for example, that 'killing a disabled infant is not morally equivalent to killing a person. Very often it is not wrong at all' (1993: 191).

28 Other principles of reason include causality (every effect has a cause), identity (each thing is separated in its existence from other things), and purpose or finality (every action has an end or purpose).

29 See Levy (2002).

30 Hans Jonas writes further:

> What is it that *society* can or cannot afford – leaving aside for the moment the question of what it has a *right* to? … The specific question seems to be

> whether society can afford to let some people die whose death might be deferred by particular means if these were authorized by society. Again, if it is merely a question of what society can or cannot afford, rather than of what it ought or ought not to do, the answer must be: Of course, it can.
>
> (1974: 115)

31 For Kant, the idea that 'life itself' could be an *object* of knowledge was impossible. Since we are mortal, all of our experience and knowledge of the world comes to us *through* human embodiment. In other words, 'life' is the condition of possibility for knowledge, meaning that knowledge of life itself would require a different perspective from the human. An objective definition of life would be available only to God, for whom mortal existence is not a precondition for knowledge. Only an immortal being could access such knowledge. Evidently this is one place where bioethics departs dramatically from the Kantian position. For bioethics there is nothing unfathomable about life because it exists objectively in nature, something that ostensibly exists independent of human consciousnesss. The definition of life generally accepted in bioethics is straightforward, so much so that at first we might wonder why Kant never thought of it; something is life 'if it is occupied with the proper care and maintenance of itself and its health' (Dean 2004: 172).

32 Significantly, while bioethics takes 'life' as a positive and empirically given *concretum*, in the biological sciences an adequate definition of life has long marked a limit and an absence (Webster and Goodwin 1982; Kull 1999; Weber and Varela 2002):

> If there is any systematic absence in the thinking about modern biology, it is a fully developed notion of an *organism*, instead of as an array of genetic and physiological processes whose unity is left unaccounted for … This absence has been underlined for decades from several sides.
>
> (Weber and Varela 2002: 101)

It is this absence that some have suggested can be filled with Jonas' idea of metabolism. See Weber and Varela (2002) for a sustained and sophisticated attempt to do just this; see also Nichols (2007) for a critique of Weber and Varela on the basis of Freud's notion of the death drive.

33 In *Bioethics in America,* Stevens (2000) also points out the role that the new criteria played in the Karen Ann Quinlan case in the US in the late 1970s.

34 Žižek references Lacan's four discourses in discussing the relationship between sovereignty and biopower. Sovereign power corresponds to the discourse of the master, while biopower corresponds to the discourse of the university. For Lacan these are not unrelated; the discourse of the university emerges out of the master discourse's ultimate failure to guarantee itself. Lacan explains the discourses in various places, but emphasises the political significance in *The Other Side of Psychoanalysis* (2007).

35 The idea of two types of desire is also reflected in recent debates among Lacanian-inspired political theorists, some of whom advocate an 'ethics of desire', and those who favour a more radical 'ethics of jouissance'. In the ethics seminar, Lacan does seem to invoke two modes of desire or at least a 'split' and 'refracted' desire (Lacan 1992: 248). Slavoj Žižek evokes this ambiguity when he asks: 'is not Lacan's entire theoretical edifice torn between these two options: between the ethics of desire/Law, of maintaining the gap, and the lethal/suicidal immersion in the Thing?' (Žižek 1997: 239).

36 The exclusion of unnatural – in this sense – desires remains. For example, Allmark writes that one point of agreement between theory-led and practice-led approaches is that one should disregard the wishes of individuals imbued with 'alien desires', which are desires imagined as externally caused or implanted. For example, with the wishes of desires suggested through hypnosis, individuals 'are not autonomous at least in relation to those desires' (Allmark 2006: 76).

37 From an article that recently appeared on *Wired Online*:

> Scientists at the Georgia Institute of Technology figured they could learn more from neuron clumps that acted more like real brains, so they've developed 'neurally controlled animats' – a few thousand rat neurons grown atop a grid of electrodes and connected to a robot body or computer-simulated virtual environment ... 'When someone is born, they're still not able to control much of their behaviour,' Wagenaar said. 'Somehow this system has to learn to control a body. Part of that comes from interactions with environment. We hope to get, at the very simple level of small nervous system, some insight into how that occurs ... ' [Steve] Potter hopes his research will eventually lead to better neural prosthetics, understanding of neural pathologies and even artificial intelligence. As for consciousness, he said, 'I don't think it will get that far. But I'd love to be proven wrong ... ' Researchers have found that lab-grown neuron cultures tend to fire in bizarrely synchronized, dish-wide waves, eerily echoing the neural patterns seen during Alzheimer's disease. 'It's possible that this is a state of arrested development,' Potter said, 'or that the networks are asleep because they're missing the parts (humans) use to wake up. It's (also) possible that the networks are in some sort of epileptic state ... ' His group has since learned to reduce the bursts with electric stimuli, which act as a massage to ease the dish-brain's stress.
>
> (Keim 2006)

38 For example, Michael Ignatieff addresses the relationship between international law and state sovereignty and acknowledges that the nature of sovereignty is changing; states are increasingly deferential to international law. He argues that a formal international rule of law system is now needed to ensure that 'the exercise of state sovereignty is conditional to some degree on observance of proper human rights behavior' (2001: 17).

39 Agamben relies on Paulus Festus' definition of the archaic legal category (see Lowrie 2007).

40 Agamben maintains that his position is in contrast to the 'authoritative current of French thought' of deconstruction (1999: 44). Deconstruction is preoccupied with the relation to the unsayable aspect of meaning to such a degree that it inaugurates only a perpetual return to the ordeal of undecidabilty, and in so doing condemns 'the handmaiden to a marriage with its theological master' (1999: 45). While deconstruction exhausts itself in 'the recognition of the presupposition of language' (i.e. the lack of ground for meaning) it forgets the original task of philosophy, which is not the revelation of presuppositions but the 'elimination' of presuppositions (1999: 45).

41 These comments written in another context convey this idea as well:

> Prison destroys me precisely when I do not fully consent to the fact that I am in prison, but maintain an inner distance towards it, stick to the illusion that 'real life is elsewhere' and indulge all the time in daydreaming about life outside, about nice things that are waiting for me after my release or

> escape … In order effectively to liberate oneself from the grip of existing social reality, one should first renounce the transgressive fantasmic supplement that attaches us to it.
>
> (Žižek 2000b: 148–49)

42 Of course, Benjamin's argument would be that lack in the context of law is covered and marked by violence. Stavrakakis is addressing his book specifically to leftists, which is probably why he focuses on the way that utopian ideals can also cover lack, although not in the law but in the identity of the subject.
43 Derrida also thought there was something about the non-conscious propulsion of a situation towards its own destruction that describes the contemporary political moment. In *Philosophy in the Time of Terror: Dialogues with Jürgen Habermas and Jacques Derrida*, (2003), Derrida characterises the contemporary political scene in terms of an auto-immune disorder. The following quote comes from 'La Raison du plus fort (y a-t-il des états voyous?)': 'The auto-immune does not only consist in harming oneself or in ruining oneself, in truth in destroying one's own [*propre*] protection … auto-immunity always threatens to deprive suicide itself of its meaning and its supposed integrity' (Derrida 2003: 71).
44 It is worth noting in the seminar on the ethics of psychoanalysis the two formulas that Lacan lays out for understanding political discourse of intellectuals:

> It was noted then that, for a long time now, there have been left-wing intellectuals and right-wing intellectuals. I would like to give you formulas for them that, however categorical they may appear at first sight, might nevertheless help to illuminate the way. The 'fool' is an innocent, a simpleton, but truths issue from his mouth that are not simply tolerated but adopted, by virtue of the fact that this 'fool' is sometimes clothed in the insignia of the jester. And in my view it is a similar happy shadow, a similar fundamental 'foolery', that accounts for the importance of the left-wing intellectual. Everyone knows that a certain way of presenting himself, which constitutes part of the ideology of the ring-wing intellectual, is precisely to play the role of what he is in fact, namely, a 'knave'. In other words, he doesn't retreat from the consequences of what is called realism; that is, when required, he admits he's a crook.
>
> (Lacan 1992: 182–83)

45 Other defence mechanisms defined by Freud are repression, regression to an earlier phase of development, and fixation on a particular stage of development. See the *Ego and the Id* (1961b).
46 The actual line from the story: 'For God's sake! – quick! – quick! – put me to sleep – or, quick! – waken me! – quick! – I say to you that I am dead!' (Poe 1845: 565).
47 In a state of exception, it is not only that life exists beyond life and death but that it is required to do so. Consider a story in *The New York Times* published 8 November 2005 that describes hunger strikes by inmates in the Guantánamo Bay detention facility:

> In recent months, many detainees have conducted hunger strikes to protest conditions and their prolonged confinement without trial, according to the accounts of lawyers who represent them. A spokesman for the military command at Guantánamo said that 27 detainees were participating in a hunger strike and that 23 of them were being fed with nasal tubes in the camp hospital. The spokesman, Lt. Col. Jeremy Martin, said the prisoners were not

> restrained for what the military calls assisted feeding. Some lawyers have filed pleadings in federal court asserting that their clients are forcibly fed and have been injured by having feeding tubes roughly inserted through their noses.

Another article on the same subject published in *The New York Times* on 9 April 2007:

> A long-term hunger strike has broken out at the American detention center at Guantánamo Bay, Cuba, with more than a dozen prisoners subjecting themselves to daily force-feeding to protest their treatment, military officials and lawyers for the detainees say … Thirteen detainees are now on hunger strikes, the largest number to endure the force-feeding regimen on an extended basis since early 2006, when the military broke a long-running strike with a new policy of strapping prisoners into restraint chairs while they are fed by plastic tubes inserted through their nostrils. The hunger strikers are now monitored so closely that they have virtually no chance to starve themselves … 'We don't have any rights here, even after your Supreme Court said we had rights,' one hunger striker, Majid al-Joudi, told a military doctor, according to medical records released recently under a federal court order … 'My wish is to die,' one reported hunger striker in the camp, Adnan Farhan Abdullatif, a 27-year old Yemeni, told his lawyer on Feb. 27, according to recently declassified notes of the meeting. 'We are living in a dying situation.'

48 A study of modernist literature discusses how it already portended to a crisis of sovereignty (Miller 2007); The work of W.B. Yeats, T.S. Eliot and Virginia Woolf indicates a lack of faith in the nation state's capacity to contend with the emerging global antagonisms and erosion of inside/outside binarisms like public/private, and global/national (Miller 2007: viii). In the work of each there is a distinct sense of this:

> The model of a strong national state … can no longer provide an adequate means of coping with a world in which, under the deterritorializing influence of capitalism and imperialism, transnational networks of trade, migration, and communication have thrown traditional forms of sovereignty into crisis.
> (Miller 2007: xiv)

49 The coherence of the concept of sovereignty has long been assumed but its waning legitimacy is also signalled by debates about the possibility of the self-negation of sovereignty, whereby a sovereign decision is made to forfeit aspects of sovereignty (such as being unlimited) (MacCormick 1999). None of these developments question sovereignty as a way to approximate justice through law.

Bibliography

Adams, Rae (1981) 'Ethical Principles in Anthropological Research: One or Many?' *Human Organization* 40(2): 155–60.

Agamben, Giorgio (1991) *Language and Death: The Place of Negativity*, Karen E. Pinkus (trans.), Minneapolis: University of Minnesota Press.

Agamben, Giorgio (1993a) *The Coming Community*, Michael Hardt (trans.), Minneapolis: University of Minnesota Press.

Agamben, Giorgio (1993b) *Infancy and History: Essays on the Destruction of Experience*, Liz Heron (trans.), New York: Verso.

Agamben, Giorgio (1996) 'Beyond Human Rights', Cesare Casarino (trans.), in *Radical Thought in Italy: A Potential Politics*, Paolo Virno and Michael Hardt (eds), Minnesota: University of Minnesota Press.

Agamben, Giorgio (1998) *Homo Sacer: Sovereign Power and Bare Life*, Daniel Heller-Roazen (trans.), Stanford: Stanford University Press.

Agamben, Giorgio (1999) *Potentialities: Collected Essays in Philosophy*, Daniel Heller-Roazen (trans. and ed.), Stanford: Stanford University Press.

Agamben, Giorgio (2000) *Means Without Ends: Notes on Politics*, V. Binette and C. Casarino (trans.), Minneapolis, University of Minnesota Press.

Agamben, Giorgio (2004) *The Open: Man and Animal*, Kevin Attell (trans.), Stanford: Stanford University Press.

Agamben, Giorgio (2005) *The State of Exception*, Kevin Attell (trans.), London: University of Chicago Press.

Allan, Trevor (2001) *Constitutional Justice*, Oxford: Oxford University Press.

Allmark, Peter (2006) 'An Argument for the Use of Aristotelian Method in Bioethics', *Medicine, Health Care and Philosophy*, 9(1): 69–79.

Altizer, Thomas (1985) *History as Apocalypse*, New York: SUNY Press.

Annas, George J. (2005) *American Bio-ethics: Crossing Human Rights and Health Law Boundaries*, New York: Oxford University Press.

Aoki, Douglas Sadao (2006) 'Letters from Lacan', *Paragraph*, 29(3): 1–20.

Aquinas, Thomas (1989 [1274]) *Summa Theologiae: A Concise Translation*, Timothy McDermott (ed.), Allen, TX: Christian Classics.

Arendt, Hannah (1976) *The Origins of Totalitarianism*, San Diego: Harcourt.

Aristotle (1962) *The Politics of Aristotle*, Ernest Barker (trans.), New York: Oxford University Press.

Aristotle (1984) *The Politics*, Carnes Lord (trans.), Chicago: University of Chicago Press.

Aristotle (1999) *The Nicomachean Ethics*, Terence Irwin (trans.), Indianapolis: Hackett Publishing.

Austin, John (1954 [1832]) *The Province of Jurisprudence Determined*, H.L. Hart (ed.), London: Weidenfeld & Nicolson.

Beauchamp, Tom and James Childress (1979) *Principles of Biomedical Ethics*, Oxford: Oxford University Press.

Benjamin, Walter (1968 [1927]) 'A Critique of Violence', in *Illuminations*, Hannah Arendt (ed.), Harry Zohn (trans.), Schoken Books: New York.

Benjamin, Walter (1978 [1916]) 'On Language as Such and the Language of Man', *Reflections*, Edmund Jephcott (trans.), New York: Harcourt Brace Jovanovich.

Bentham, Jeremy (1977 [1776]) 'A Fragment on Government', in *A Comment on the Commentaries and A Fragment on Government*, J.H. Burns and H.L. Hart (eds), London: Athlone Press.

Besson, Samantha (2006) 'Sovereignty on Conflict', in *Towards an International Legal Community: The Sovereignty and States and the Sovereignty of International Law*, C. Warbrick and S. Tierney (eds), London: The British Institute of International and Comparative Law.

Borochowitz, Dalit (2004) 'Reflections on the Researcher–Participant Relationship and the Ethics of Dialogue', *Ethics & Behavior*, 12(2): 175–86.

Bosk, Charles (2007) 'The New Bureaucracies of Virtue or When Form Fails to Follow Function', *Political and Legal Anthropology Review*, 30(2): 192–209.

Brainard, Jeffrey (2001) 'The Wrong Rules for Social Science? Scholars Say US Human Subjects Protections Designed for Medical Studies Hinder Vital Work', *Chronicle of Higher Education*, 8 March: A21–A23.

Braybrooke, David (2001) *Natural Law Modernized*, Toronto: University of Toronto Press.

Butler, Judith (1993) *Bodies that Matter: On the Discursive Limits of Sex*, New York: Routledge.

Butler, Judith (1997) *The Psychic Life of Power: Essays in Subjection*, Stanford: Stanford University Press.

Butler, Judith (2000a) *Antigone's Claim: Kinship between Life and Death*, New York: Columbia University Press.

Butler, Judith (2000b) 'Competing Universalities', in Judith Butler, Ernesto Laclau and Slavoj Žižek, *Contingency, Hegemony, Universality: Contemporary Dialogues on the Left*, London: Verso.

Butler, Judith (2002) 'Is Kinship Always Already Heterosexual?' *Differences: A Journal of Feminist Cultural Studies*, 13(1): 14–44.

Butler, Judith (2004) *Precarious Life: The Powers of Mourning and Violence*, New York: Verso.

Caplan, Paul (ed.) (2003) *The Ethics of Anthropology: Debates and Dilemmas*, London: Routledge.

Capps, Patrick (2006) 'Sovereignty and the Identity of Legal Orders', in *Towards an International Legal Community: The Sovereignty and States and the Sovereignty of International Law*, C. Warbrick and S. Tierney (eds), London: The British Institute of International and Comparative Law.

Cascais, Fernando (2003) 'Bioethics: A Tentative Balance', *Studia Bioetica*, 1(1): 25–34.

Cassell, Joan (1980) 'Ethical Principles for Conducting Fieldwork', *American Anthropological Association*, 82(1): 28–41.

Caton, Steven (2006) 'Coetzee, Agamben and the Passion of Abu Griab', *American Anthropologist*, 108(1): 114–23.

Caygill, Howard (1995) *A Kant Dictionary*, Cambridge: Blackwell.

Christakis, Nicholas. (1992) 'Ethics are Local: Engaging Cross-cultural Variation in the Ethics of Clinical Research', *Social Science and Medicine*, 35(9): 1079–91.

Clarke, Adele E., Jennifer Fishman, Jennifer Fosket, Laura Mamo and Janet Shim (2003) 'Biomedicalization: Technoscientific Transformations of Health, Illness, and US Biomedicine', *American Sociological Review*, 68(2): 161–94.

Cohen, Carl I. (1993) 'The Biomedicalization of Psychiatry: A Critical Review', *Community Mental Health Journal*, 29(6): 509–21.

Cook, Robert Manuel (1964) *Niobe and her Children*, Cambridge: Cambridge University Press.

Copjec, Joan (1994) *Read My Desire: Lacan Against the Historicists*, Cambridge, MA: MIT Press.

Copjec, Joan (ed.) (1996) *Radical Evil*, New York: Verso.

Copjec, Joan (2002) *Imagine There's No Woman*, Cambridge, MA: MIT Press.

Corrigan, Richard and Derek Sayer (1985) *The Great Arch: English State Formation as a Cultural Revolution*, New York: Routledge.

Davies, Margaret (1996) *Delimiting the Law: Postmodernism and the Politics of Law*, Chicago: Pluto Press.

Dean, Mitchell (1994) *Critical and Effective Histories: Foucault's Methods and Historical Sociology*, New York: Routledge.

Dean, Mitchell (2004) 'Four Theses on the Powers of Life and Death', *Contretemps*, 5: 16–29; available online at www.usyd.edu.au/contretemps/5december2004/dean.pdf (accessed 6 November 2008).

Derrida, Jacques (1976) *Of Grammatology*, Gayatri C. Spivak (trans.), Baltimore and London: Johns Hopkins University Press.

Derrida, Jacques (1986) *Glas*, John P. Leavey and Richard Rand (trans.). Lincoln: University of Nebraska Press.

Derrida, Jacques (1987) *The Post Card*, Alan Bass (trans.), Chicago: University of Chicago Press.

Derrida, Jacques (1992) 'Force of Law: The Mystical Foundation of Authority', in *Deconstruction and the Possibility of Justice*, Drucilla Cornell, Michel Rosenfeld and David Gray Carlson (eds), New York: Routledge.

Derrida, Jacques (2000) *Of Hospitality: Anne Dufourmantelle Invites Jacques Derrida to Respond*, Stanford: Stanford University Press.

Derrida, Jacques (2003) *Philosophy in the Time of Terror: Dialogues with Jürgen Habermas and Jacques Derrida*, Giovanna Borradori (ed.), Chicago: Chicago University Press.

Docherty, Thomas (2006) *Aesthetic Democracy*, Stanford: Stanford University Press.

Douzinas, Costas and Ronnie Warrington (1994) *Justice Miscarried: Ethics, Aesthetic and the Law*, Toronto: Harvester Wheatsheaf.

Draper, G.I.A.D. (1990) 'Grotius' Place in the Development of Legal Ideas about War', *Hugo Grotius and International Relations*, Hedley Bull, Benedict Kingsbury and Adam Roberts (eds), New York: Oxford University Press.

Estes, Carroll. L. and Elizabeth A. Binney (1989) 'The Biomedicalization of Aging: Dangers and Dilemmas', *Gerontologist*, 29(5): 587–96.

Evans, Dylan (1996) *An Introductory Dictionary of Lacanian Psychoanalysis*, New York: Routledge.

Finnemore, Martha (2003) *The Purpose of Intervention: Changing Beliefs about the Use of Force*, Ithaca: Cornell University Press.

Finnis, John (1980) *Natural Law and Natural Rights*, Oxford: Clarendon Press.

Finnis, John (1996) 'The Truth in Legal Positivism', in *The Autonomy of Law: Essays in Legal Positivism*, Robert P. George (ed.), New York: Oxford University Press.

Fitzpatrick, Peter (2000) *Modernism and the Grounds of Law*, Cambridge: Cambridge University Press.

Foucault, Michel (1974) *The Order of Things: An Archaeology of the Human Sciences*, London: Routledge.

Foucault, Michel (1977) *Discipline and Punish: The Birth of the Prison*, New York: Vintage.

Foucault, Michel (1978) *The History of Sexuality, Vol. 1: An Introduction*, Robert Hurley (trans.), London: Allen Lane.

Foucault, Michel (1980) *Michel Foucault: Power/Knowledge*, Colin Gordon (ed.), New York: Harvester.

Foucault, Michel (1988) 'The Political Technology of Individuals', in *Technologies of the Self: A Seminar with Michel Foucault*, Luther H. Martin, Huck Gutman and Patrick H. Hutton (eds), London: Tavistock Press.

Foucault, Michel (1991) 'Governmentality', in *The Foucault Effect: Studies in Governmentality*, Graham Burchell, Colin Gordon and Peter Miller (eds), Hemel Hempstead: Harvester Wheatsheaf, pp. 87–104.

Foucault, Michel (1996) 'A Preface to Transgression … ', in *Language, Counter-memory, Practice: Selected Essays and Interviews*, Donald F. Bouchard (ed.), Ithaca: Cornell University Press.

Foucault, Michel (2000) *Power: Essential Works of Michel Foucault 1954–1984, Vol. 3*, James D. Faubion (ed.), New York: New Press.

Foucault, Michel (2004) *Society Must Be Defended*, David Macey (trans.), New York: Picador.

Foucault, Michel (2007) *Security, Territory, Population: Lectures at the Collège de France, 1977–78*, G. Burchell (trans.), Basingstoke and New York: Palgrave Macmillan.

Freud, Sigmund (1961a [1920]) *Beyond the Pleasure Principle*, in *The Standard Edition of the Complete Psychological Works of Sigmund Freud*, Vol. 18, James Strachey (ed.), London: Hogarth Press.

Freud, Sigmund (1961b [1927]) *The Ego and the Id*, in *The Standard Edition of the Complete Psychological Works of Sigmund Freud*, Vol. 19, James Strachey (ed.), London: Hogarth Press.

Freud, Sigmund, (1989 [1930]) *Civilization and Its Discontents*, New York: W.W. Norton.

Fuller, Lon (1969) *The Morality of Law*, second edition, New Haven, CT: Yale University Press.

Galgay, Erin. (2007) 'The Wisdom of Humility: A Discussion of Responsibility and Bioethics', in *Ethos: Boston College Student Bioethics Research Journal*, 2; available online at www.bc.edu/clubs/mendel/ethos/archives/2007/galgay.shtml (accessed 6 November 2008).

George, Robert P. (1996) 'Natural Law and Positive Law', in *The Autonomy of Law: Essays on Legal Positivism*, Robert P. George (ed.), New York: Oxford University Press.

Gerlach, Judy (2002) 'Bioethics Inside the Beltway: What Should IRBs Consider when Applying the Privacy Rule to Research?' *Kennedy Institute of Ethics Journal*, 12(3): 299–303.

Gert, Bernard, Charles M. Culver and K. Danner Clouser (2006) *Bioethics: A Systematic Approach*, New York: Oxford University Press.

Goldsmith, Jack A. and Eric Posner (2005) *The Limits of International Law*, New York: Oxford University Press.

Gordis, Robert (1961) 'Natural Law and Religion', in *Natural Law and Modern Society*, John Cogley (ed.), New York: World Press.

Gordon, Elisa (2000) 'When Oral Consent Will Do', *Field Methods*, 12(3): 235–38.

Gordon, Elisa (2003) 'Trials and Tribulations of Navigating IRBs: Anthropological and Biomedical Perspectives of "Risk" in Conducting Human Subjects Research', *Anthropological Quarterly*, 76(2): 299–320.

Green, Leslie (2003) 'Legal Positivism', in *The Stanford Encyclopedia of Philosophy (Spring 2003 Edition)*, Edward N. Zalta (ed.); available online at http://plato.stanford.edu/archives/spr2003/entries/legal-positivism/ (accessed 6 November 2008).

Grigg, Russell (2001) 'Absolute Freedom and Major Structural Change', *Paragraph*, 24(2): 111–25.

Haggerty, Kevin (2004) 'Ethics Creep: Governing Social Science Research in the Name of Ethics', *Qualitative Sociology*, 27(4): 391–414.

Halse, Christine and Anne Honey (2005) 'Unravelling Ethics: Illuminating the Moral Dilemmas of Research Ethics', *Signs*, 30(4): 2141–63.

Hare, R.M. (1975) 'Abortion and the Golden Rule', *Philosophy and Public Affairs*, 4 (3): 201–22.

Hart, H.L.A. (1958) 'Positivism and the Separation of Law and Morals', *Harvard Law Review*, 71(4): 593–629.

Hart, H.L.A. (1961) *The Concept of Law*, Oxford: Oxford University Press.

Hart, H.L.A. (1965) 'Book Review of *The Morality of Law*', *Harvard Law Review*, 78 (6): 1281–96.

Hart, H.L.A. (1994) *The Concept of Law*, second edition, Oxford: Oxford University Press.

Hegel, G.W.F. (1977) *The Phenomenology of Spirit*, A.V. Miller (trans.), Oxford: Oxford University Press.

Hogue, Michael S. (2007) 'Theological Ethics and Technological Culture: A Biocultural Approach', *Zygon*, 42(1): 77–95.

Homan, Roger (1992) 'The Ethics of Open Methods', *British Journal of Sociology*, 43 (3): 321–32.

Hyland, William F. and David S. Baime (1976) '*In Re Quinlan*: A Synthesis of Law and Medical Technology', *Rutgers Camden Law Journal*, 8(1): 37–64.

Ignatieff, Michael (2001) *Human Rights Politics and Ideology*, Amy Gutman (ed.), Princeton: Princeton University Press.

jagodzinski, jan (2004) *Youth Fantasies: The Perverse Landscape of the Media*, New York: Palgrave Macmillan.

James, Alan (1999) 'The Practice of Sovereign Statehood in Contemporary International Society', *Political Studies*, 47(3): 460–62.

Johnston, Douglas M. (1998) 'The Heritage of Political Thought in International Law,' in *The Structure and Process of International Law: Essays in Legal*

Philosophy, Doctrine and Theory, Ronald St J. Macdonald and Douglas M. Johnston (eds), Lancaster: Martinus Nijhoff.

Jonas, Hans (1969) 'Philosophical Reflections on Experimenting with Human Subjects', *Daedalus*, 98(2): 219–47.

Jonas, Hans (1974) 'Philosophical Reflections on Experimenting with Human Subjects', in *Philosophical Essays*, New Jersey: Prentice-Hall.

Jonas, Hans (1984) *The Imperative of Responsibility*, Hans Jonas and David Herr (trans.), Chicago: University of Chicago Press.

Jonas, Hans (2001) *The Phenomenon of Life: Toward a Philosophical Biology*, Evanston, IL: Northwestern University Press.

Jonsen, Albert (1998) *The Birth of Bioethics*, New York: Oxford University Press.

Journal of the American Medical Association (1968) 'What and When is Death?' (editorial), 204(6): 219–20.

Kagawa, C. (2009) 'Neuroethics and Bioethics: Implication of Balkanization Controversy', *Brain Nerve*, 61(1): 11–17.

Kainz, Howard P. (2004) *Natural Law: An Introduction and Re-examination*, Chicago: Open Court Publishing.

Kant, Immanuel (1964) *Groundwork for the Metaphysic of Morals*, New York: Paton, Harper & Row.

Kant, Immanuel (1982) *Critique of Pure Reason*, W. Schwarz (trans.), Germany: Scientia Verlag Aalen.

Kant, Immanuel (1987) *Critique of Judgment*, Werner Pluhar (trans.) Indianapolis: Hackett.

Kant, Immanuel (1993) *Critique of Practical Reason*, Lewis White Beck (trans.), New York: Macmillan.

Kant, Immanuel (1998) *Religion within the Boundaries of Mere Reason*, Allan Wood and G. Di Giovanni (trans.), Cambridge: Cambridge University Press.

Kant, Immanuel (2001) *The Critique of the Power of Judgment*, P. Guyer and E. Matthews (trans.), New York: Cambridge University Press.

Keating, Michael (2001) *Plurinational Democracy: Stateless Nations in a Post-Sovereignty Era*, Oxford: Oxford University Press.

Keenan, Dennis King (2005) *The Question of Sacrifice*, Bloomington: Indiana University Press.

Keim, Brandon (2006) 'It's Alive(ish)', *Wired Online*, 8 August 2006; available at www.wired.com/medtech/health/news/2006/08/71457 (accessed 3 August 2010).

Kellogg, Catherine (2003) 'Mourning the Law', *Philosophy and Social Criticism*, 29 (4): 361–75.

Kelsen, Hans (1945) *General Theory of Law and State*, A. Wedberg (trans.), Cambridge, MA: Harvard University Press.

Kitcher, Patricia (1996) 'Introduction', in Immanuel Kant, *Critique of Pure Reason*, W. Pluhar (trans.), Indianapolis: Hackett.

Klein, Allison (2008) 'DC Police to Check Drivers in Violence-Plagued Trinidad', *Washington Post*, 5 June: A1.

Krell, David Farrell (2005) *The Tragic Absolute: German Idealism and the Languishing God*, Bloomington and Indianapolis: Indiana University Press.

Krishna, Sankaran (1999) *Postcolonial Insecurity: India, Sri Lanka and the Question of Nationhood*, Minneapolis: University of Minnesota Press.

Kull, Kalevi (1999) 'Biosemiotics in the Twentieth Century: A View from Biology', *Semiotica*, 127(1/4): 385–414.

Lacan, Jacques (1988) *Book II: The Ego in Freud's Theory and in the Technique of Psychoanalysis*, Jacques Alain-Miller (ed.), Sylvana Tomaselli (trans.), with notes by John Forrester, New York: W.W. Norton.

Lacan, Jacques (1992 [1959–60]) *Book VII: The Ethics of Psychoanalysis 1959–60*, Jacques-Alain Miller (ed.), Daniel Porter (trans.), London: Routledge.

Lacan, Jacques (1993) *Book III: The Psychoses 1955–56*, Jacques-Alain Miller (ed.), R Grigg (trans.), London: Routledge.

Lacan, Jacques (1998a) *Book XI: The Four Fundamental Concepts of Psychoanalysis*, Jacques-Alain Miller (ed.), Alan Sheridan (trans.), New York: W.W. Norton.

Lacan, Jacques (1998b) *Book XX: Encore, On Feminine Sexuality: The Limits of Love and Knowledge*, Bruce Fink (trans.), with notes by Bruce Fink, New York: W.W. Norton.

Lacan, Jacques (2006) *Ecrits: A Selection*. Bruce Fink (trans.), New York: W.W. Norton.

Lacan, Jacques (2007) *Book XVII: The Other Side of Psychoanalysis*, Russell Grigg (trans.), New York: W.W. Norton.

Lacey, Nicola (2002) 'Violence, Ethics and Law', in *Visible Women: Essays on Feminist Legal Theory and Political Philosophy*, Susan James and Stephanie Palmer (eds), Oxford: Hart Publishing.

Lederman, Rena (2007) 'Comparative "Research": A Modest Proposal for the Object of Ethics Regulation', *Political and Legal Anthropology Review*, 30(2): 305–27.

Lesky, Albin, James Willis and Cornelius de Heer (1996) *A History of Greek Literature*, London: Gerald Duckworth.

Levine, Michael P. (1995) 'The Last Word', *International Journal of Eating Disorders*, 3(1): 92–95.

Levy, David J. (2002) *Hans Jonas: The Integrity of Thinking*, Missouri: University of Missouri Press.

Lindgren, James (2002) 'Anticipating Problems –32 Doing Social Science Research in the Shadow of the Law', *Sociological Methodology*, 32(29): 29–32.

Lindseth, Peter (1999) 'Democratic Legitimacy and the Administrative Character of Supranationalism: The Example of the European Community', *Columbia Law Review*, 99(3): 628–738.

Lippert, Randy (2005) *Sanctuary, Sovereignty and Sacrifice: Canadian Sanctuary Incidents, Power, and Law*, Vancouver: UBC Press.

Lipson, Juliene G. (1994) 'Ethical Issues in Ethnography', in *Critical Issues in Qualitative Research Methods*, J. Morse (ed.), Thousand Oaks, CA: Sage.

Lock, Margaret (2001) 'The Alienation of Body and Tissue and the Biopolitics of Immortalized Cell Lines', *Body and Society*, 7(2/3): 63–91.

Lowrie, Michele (2007) 'Sovereignty Before the Law: Agamben and the Roman Republic,' *Law and Humanities*, 1(1): 31–55.

MacCormick, Eric (1996) 'The Concept of Law and *The Concept of Law*', in *The Autonomy of Law: Essay in Legal Positivism*, Robert P. George (ed.), New York: Oxford University Press.

MacCormick, Eric (1999) *Questioning Sovereignty: Law, State and the Nation in the European Community*, Oxford: Oxford University Press.

Marshall, Patricia (2003) 'Human Subject Protections, Institutional Review Boards, and Cultural Anthropological Research', *Anthropological Quarterly*, 76(2): 269–85.

Mattingly, Cheryl (2005) 'Toward a Vulnerable Ethics of Research Practice', *Health: An Interdisciplinary Journal for the Social Study of Health, Illness and Medicine*, 9(4): 453–71.

Mauthner, Maureen (2000) 'Snippets and Silences: Ethics and Reflexivity in Narratives of Sistering', *Social Research Methodology*, 3(4): 287–306.

May, Larry (2005) *Crimes Against Humanity: A Normative Account*, Cambridge: Cambridge University Press.

Mayhew, Robert (1997) *Aristotle's Criticism of Plato's Republic*, Lanham, MD: Rowman & Littlefield.

Melville, Herman (1997 [1853]) 'Bartleby the Scrivener', in *The Complete Shorter Fiction*, New York: Random House.

Milde, Michael (2002) 'Contemporary State Sovereignty under the Microscope', *The University of Toronto Law Journal*, 52(2): 275–301.

Miller, Andrew John (2007) *Modernism and the Crisis of Sovereignty*, New York: Routledge.

Milovanovic, Dragan (2003) *An Introduction to the Sociology of Law*, New York: Willow Tree Press.

Murphy, Mark C. (2006) *Natural Law in Jurisprudence and Politics*, New York: Cambridge University Press.

Nef, George (2002) 'Globalization and the Crisis of Sovereignty, Legitimacy, and Democracy', *Latin American Perspectives*, 29(6): 59–69.

Negri, Antonio (2003) 'The Ripe Fruit of Redemption', 26 July 2003; available online at www.generation-online.org/t/negriagamben.htm (accessed 6 November 2008).

Nelsen, Judith C. (1994) 'Ethics, Gender, and Ethnicity in Single-case Research Evaluation', *Journal of Social Science Research*, 18(3/4): 139–52.

Nichols, Joshua (2007) 'Beyond the Autopoetic Principle? A Preliminary Analysis of the Freudian Cell', *Body and Society*, 13(1): 21–40.

Norris, Andrew (2005) 'Giorgio Agamben and the Politics of the Living Dead', in *Politics, Metaphysics and Death: Essays on Giorgio Agamben's* Homo Sacer, Andrew Norris (ed.), London: Duke University Press.

Nussbaum, Martha (1986) *The Fragility of Goodness: Luck and Ethics in Greek Tragedy and Philosophy*, New York: Cambridge University Press.

Oakes, Michael J. (2002) 'Risks and Wrong in Social Science Research An Evaluator's Guide to IRBs', *Evaluation Review*, 26(5): 443–79.

Ojakangas, Mike (2005) 'Impossible Dialogue on Power: Agamben and Foucault', *Foucault Studies*, 2: 5–28.

Oliver, Peter C. (2003) 'Sovereignty in the Twenty-First Century', *King's College Law Journal*, 14(2): 137–46.

O'Neill, Onara (2002) *Autonomy and Trust in Bioethics (Gifford Lectures 2001)*, Cambridge: Cambridge University Press.

Painter, Joe (2002) 'State and Governance', in *A Companion to Economic Geography*, Eric Sheppard and Trevor J. Barnes (eds), Baltimore: Blackwell.

Parens, Erik, Josephine Johnston and Jacob Moses (2008) 'Do we Need "Synthetic Bioethics"?' *Science*, 321(5895): 1449.

Passavant, Paul (2006) 'The Contradictory State of Giorgio Agamben', *Political Theory*, 35(22): 147–74.

Pavkovic, Aleksander and Peter Radan (2003) 'In Pursuit of Sovereignty and Self-Determination: Peoples, States and Secession in the International Order', *Macquarie Law Journal*, 3: 1–12.

Pavlich, George (2005) *Governing Paradoxes of Restorative Justice*, London: Glasshouse Press.

Pavlich, George (2007) 'Ethics, Universal Principles and Restorative Justice', in *Handbook of Restorative Justice*, Gerry Johnston and Daniel W. Van Ness (eds), Portland: Willan Publishing.

Peckman, Steven. (2001) 'Local Institutional Review Boards', in *Ethical and Policy Issues in Research Involving Human Participants*, NBAC (ed.), Bethesda: National Bioethics Advisory Commission.

Plattner, Stuart (2003) 'Human Subjects, Protection and Cultural Anthropology', *Anthropological Quarterly*, 76(2): 287–97.

Poe, Allen Edgar (1845) 'The Facts in the Case of M. Valdemar', *The American Review: A Whig Journal*, 2(1): 561–65.

Potter, Van Rensselaer (1988) *Global Ethics: Building on the Leopold Legacy*, Michigan: Michigan State University Press.

Rancière, Jacques (2004) 'Who is the Subject of Human Rights?' *South Atlantic Quarterly*, 103(2–3): 297–310.

Rapp, R. (2003) 'Cell Life and Death, Child Life and Death: Genomic Horizons, Genetic Diseases, Family Stories', in *Remaking Life and Death. Toward an Anthropology of the Biosciences*, S. Franklin and M. Lock (eds), Santa Fe: School of American Research Press.

Raz, Joseph (1979) *The Authority of Law*, Oxford: Clarendon Press.

Raz, Joseph (1990) *Practical Reason and Norms*, Oxford: Clarendon Press.

Restuccia, Frances (2006) *Amorous Acts: Lacanian Ethics in Modernism, Film and Queer Theory*, Stanford: Stanford University Press.

Rose, Nicholas (1999) *Powers of Freedom Reframing Political Thought*, Cambridge: Cambridge University Press.

Ross, Alison, (2007) 'Introduction', *South Atlantic Quarterly*, 107(1): 1–13.

Rothbard, Murray N. (1983) *The Ethics of Liberty*, Atlantic Heights, NJ: Humanities Press.

Rothman, David J. (1991) *Strangers at the Bedside: A History of How Law and Bioethics Transformed Medical Decision Making*, New York: Walter de Gruyter (second edition 1993).

Saurette, Paul (2005) *The Kantian Imperative: Humiliation, Common Sense, Politics*, Toronto: University of Toronto Press.

Scavone, Robert (1985) 'Natural Law, Obligation, and the Common Good: What Finnis Can't Tell Us', *University of Toronto Faculty of Law Review*, 43(1): 90–115.

Schauer, Frederick (1996) 'Positivism as Pariah', in *The Autonomy of Law: Essays on Legal Positivism*, Robert P. George (ed.), New York: Oxford University Press.

Scheper-Hughes, Nancy (1995) 'The Primacy of the Ethical: Propositions for a Militant Anthropology', *Current Anthropology*, 36(3): 409–20.

Scheper-Hughes, Nancy (2001) 'Commodity Fetishism in Organ Trafficking', *Body and Society*, 7(2): 31–62.

Scheper-Hughes, Nancy (2002) *Commodifying Bodies*, London: Sage.

Schirmacher, Wolfgang (1987) '*Homo Generator*: The Challenges of Gene Technology', in *Technology and Responsibility*, Paul T. Durbin (ed.), Amsterdam: D. Reidel Publishing.

Schmitt, Carl (1985 [1934]) *Political Theology: Four Chapters on the Concept of Sovereignty*, George D. Schwab (trans.) Cambridge, MA: MIT Press.

Shannon, Jennifer (2007) 'Informed Consent: Documenting the Intersection of Bureaucratic Regulation and Ethnographic Practice', *Political and Legal Anthropology Review*, 30(2): 229–48.

Shaw, Devin Z. (2005) 'The Step Outside: The Act, Democracy and Its Discontents', *Critical Sense*, 7(1): 145–53.

Shea, Christopher (2000) 'Don't Talk to the Humans: The Crackdown on Social Science Research', *Lingua Franca*, 10(6): 26–34.

Sihvola, Juha (2002) 'Aristotle on Sex and Law', in *The Sleep of Reason: Erotic Experience and Sexual Ethics in Ancient Greece and Rome*, Martha Nussbaum and Juha Sihvola (eds), Chicago: University of Chicago Press.

Singer, Peter (1993) *Practical Ethics*, second edition, Cambridge: Cambridge University Press.

Smith, Robin (1999) 'Dialectic and Method in Aristotle', in *From Puzzles to Principles? Essays on Aristotle's Dialectic*, May Sim (ed.), Lanham, MA: Lexington Books.

Sophocles (1974) *The Theban Plays*, New York: Penguin Books.

Sorel, Georges (1999 [1915]) *Reflections on Violence*, Cambridge: Cambridge University Press.

Starzl, Thomas (1992) *The Puzzle People: Memoirs of a Transplant Surgeon*, Pittsburgh: University of Pittsburgh Press.

Stavrakakis, Yannis (2005) 'The Lure of Antigone', in *Traversing the Fantasy: Critical Responses to Slavoj Žižek*, Geoff Boucher and Matthew Sharpe (eds), Burlington: Ashgate.

Stavrakakis, Yannis (2007) *The Lacanian Left: Psychoanalysis, Theory and Politics*, New York: SUNY Press.

Stevens, Tina (2000) *Bioethics in America: Origins and Cultural Politics*, Baltimore: Johns Hopkins University Press.

Summer, L.W. and Joseph Boyle (1996) *Philosophical Perspective on Bioethics*, Toronto: University of Toronto Press.

Susanne, Charles, Maria Casado and María Jesús Buxo (2005) 'What Challenges Offers Nanotechnology to Bioethics?' *Law and the Human Genome Review*, 22: 27–45.

Tadros, Victor (1998) 'Between Governance and Discipline: The Law and Michel Foucault', *Oxford Journal of Legal Studies*, 18(1): 75–103.

Tierney, Stephen (2006) 'Introduction', in *Towards an International Legal Community: The Sovereignty and States and the Sovereignty of International Law*, Colin Warbrick and Stephen Tierney (eds), London: The British Institute of International and Comparative Law.

Toumlin, Stephen (2007) 'How Medicine Saved the Life of Ethics', in *Bioethics: An Introduction to the History, Methods and Practice*, Nancy S. Jecker, Albert R. Jonson and Robert A. Pearlman (eds), Mississauga: Jones and Bartlet Publishers.

Tuhkanen, Mikko (2003) 'Review of *Antigone's Claim*, by Judith Butler', *UMBR(a): A Journal of the Unconscious*, 1: 140–44.

Turner, Terence (2003) 'Class Projects, Social Consciousness and the Contradictions of "Globalization"', in *Globalization, the State and Violence*, Jonathan Freidman (ed.), California: Alta Mira Press.

United Nations Office of the High Commissioner for Human Rights (2006) *The Core International Human Rights Treaties*, New York: United Nations.

Van den Hoonaard, William (2001) 'Is Research Ethics Review a Moral Panic?' *Canadian Review of Sociology and Anthropology*, 38(1): 19–36.

Van den Hoonaard, William (ed.) (2002) *Walking the Tightrope: Ethical Issues for Qualitative Researchers*, Toronto: University of Toronto Press.

Veitch, Scott, Emilios Christodoulis and Lindsey Farmer (2007) *Jurisprudence: Themes and Concepts*, New York: Routledge-Cavendish.

Vernant, Jean-Pierre and Pierre Vidal-Naquet (1988) *Myth and Tragedy in Ancient Greece*, New York: Zone Books.

Vogel, Lawrence (1996) '"The Outcry of Mute Things": Hans Jonas' Imperative of Responsibility', in *Minding Nature: The Philosophers of Ecology*, David Macaulay (ed.), New York: Guilford Press.

Vogel, Lawrence, (2006) 'Natural Law Judaism? The Genesis of Bioethics in Hans Jonas, Leo Strauss, and Leon Kass', *Hastings Center Report*, 6(3): 32–44.

Waldorn, Jeremy (1991) 'The Irrelevance of Moral Objectivity', in *Natural Law Theory*, Robert P. George (ed.), New York: Oxford University Press.

Waldorn, Jeremy (2006) 'Are Constitutional Norms Legal Norms?' *Fordham Law Review*, 75(3): 1697–713.

Walsh, Lisa (1999) 'Her Mother Herself: The Ethics of the Antigone Family Romance', *Hypatia,* 14(3): 96–125.

Wax, Murray (1995) 'Informed Consent in Applied Research: A Comment', *Human Organisation,* 54(3): 330–31.

Weber, Andreas and Francisco Varela (2002) 'Life after Kant: Natural Purposes and the Antipoetic Foundations of Biological Individuality', *Phenomenology and the Cognitive Sciences*, 1(2): 97–125.

Webster, Gary and Brian C. Goodwin (1982) 'The Origin of Species: A Structuralist Approach', *Journal of Social and Biological Structure*, 5(1): 15–47.

Whyte, Jessica (2009) '"I Would Prefer Not To": Giorgio Agamben, Bartleby and the Potentiality of Law', *Law and Critique*, 20(3): 309–24.

Wolpe, Paul R. (2003) 'Neurotechnology, Cyborgs, and the Sense of Self', in *Neuroethics: Mapping the Field*, S.J. Marcus (ed.), New York: Dana Press.

Žižek, Slavoj (1989) *The Sublime Object of Ideology*, New York: Verso.

Žižek, Slavoj (1993) *Tarrying with the Negative*, Durham: Duke University Press.

Žižek, Slavoj (1997) *The Plague of Fantasies (Wo es War)*, New York: Verso.

Žižek, Slavoj (1999) *Ticklish Subject: The Absent Centre of Political Ontology*, London: Verso.

Žižek, Slavoj (2000) *The Fragile Absolute: Or, Why is the Christian Legacy Worth Fighting For?* London and New York: Verso.

Žižek, Slavoj (2001) *On Belief (Thinking in Action)*, New York: Routledge.

Žižek, Slavoj (2002a) *For They Know Not What They Do*, second edition, London: Routledge.

Žižek, Slavoj (2002b) *Revolution at the Gates: Žižek on Lenin*, New York: Verso.

Žižek, Slavoj (2003) 'A Symptom of What?' *Critical Inquiry*, 29(3): 486–503.

Žižek, Slavoj (2004a) *Organs Without Bodies: On Deleuze and Consequences*, New York: Routledge.

Žižek, Slavoj (2004b) 'From Politics to Biopolitics … and Back', *South Atlantic Quarterly*, 103(2/3): 501–21.

Žižek, Slavoj (2004c) 'The Structure of Domination Today: A Lacanian Perspective', *Studies in Eastern European Thought*, 54: 383–403.

Žižek, Slavoj (2004d) *Iraq the Borrowed Kettle*, New York: Verso.
Žižek, Slavoj (2006) *The Parallax View*, Cambridge, MA: MIT Press.
Žižek, Slavoj (2007) 'Divine Violence and Liberated Territories: *Soft Targets* Talks to Slavoj Žižek', *Soft Targets*, 17 March; available online at www.softtargetsjournal.com/web/Žižek.php (accessed 6 November 2008).
Zupancic, Alenka (2000) *Ethics of the Real: Kant, Lacan*, New York: Verso.

Index

Printed in the USA/Agawam, MA
October 16, 2012

569754.029